# SpringerBriefs in Computer Science

For further volumes:
http://www.springer.com/series/10028

Annalisa Appice · Anna Ciampi
Fabio Fumarola · Donato Malerba

# Data Mining Techniques in Sensor Networks

## Summarization, Interpolation and Surveillance

 Springer

Annalisa Appice
Anna Ciampi
Fabio Fumarola
Donato Malerba
Dipartimento di Informatica
Università degli Studi di Bari "Aldo Moro"
Bari
Italy

ISSN 2191-5768          ISSN 2191-5776  (electronic)
ISBN 978-1-4471-5453-2   ISBN 978-1-4471-5454-9  (eBook)
DOI 10.1007/978-1-4471-5454-9
Springer London Heidelberg New York Dordrecht

Library of Congress Control Number: 2013944777

Printed on acid-free paper

Springer is part of Springer Science+Business Media (www.springer.com)

# Preface

## Preamble

Sensor networks consist of distributed devices, which monitor an environment by collecting data (light, temperature, humidity,…). Each node in a sensor network can be imagined as a small computer, equipped with the basic capacity to sense, process, and act. Sensors act in dynamic environments, often under adverse conditions.

Typical applications of sensor networks include monitoring, tracking, and controlling. Some of the specific applications are photovoltaic plant controlling, habitat monitoring, traffic monitoring, and ecological surveillance. In these applications, a sensor network is scattered in a (possibly large) region where it is meant to collect data through its sensor nodes.

While the technical problems associated with sensor networks have reached certain stability, managing sensor data brings numerous computational challenges [1, 5] in the context of data collection, storage, and mining. In particular, learning from data produced from a sensor network poses several issues: sensors are distributed; they produce a continuous flow of data, eventually at high speeds; they act in dynamic, time-changing environments; the number of sensors can be very large and dynamic. These issues require the design of efficient techniques for processing data produced by sensor networks. These algorithms need to be executed in one step of the data, since typically it is not always possible to store the entire dataset, because of storage and other constraints.

Processing sensor data has developed new software paradigms, both creating new techniques or adapting, for network computing, old algorithms of earlier computing ages [2, 3]. The traditional knowledge discovery environment has been adapted to process data streams generated from sensor networks in (near) real time, to raise possible alarms, or to supplement missing data [6]. Consequently, the development of sensor networks is now accompanied by several algorithms for data mining which are modified versions of clustering, regression, and anomaly detection techniques from the field of multidimensional data series analysis in other scientific fields [4].

The focus of this book is to provide the reader with an idea of *data mining techniques in sensor networks*. We have taken special care to illustrate the impact of data mining in several network applications by addressing common problems, such as data summarization, interpolation, and surveillance.

## Book Organization

The book consists of five chapters.

Chapter 1 provides an overview of sensor networks. Since the book is concerned with data mining in sensor networks, overviews of sensor networks and data streams, produced by sensor networks, are provided in this part. We give an overview of the most promising streaming models, which can be embedded in intelligent sensor network platforms and used to mine real-time data for a variety of analytical insights.

Chapter 2 is concerned with summarization in sensor networks. We provide a detailed description with experiments of a clustering technique to summarize data and permit the storage and querying of this amount of data, produced by a sensor network in a server with limited memory. Clustering is performed by accounting for both spatial and temporal information of sensor data. This permits the appropriate trade-off between size and accuracy of summarized data. Data are processed in windows. Trend clusters are discovered as a summary of each window. They are clusters of georeferenced data, which vary according to a similar trend along the time horizon of the window. Data warehousing operators are introduced to permit the exploration of trend-clustered data from coarse-grained and inner-grained views of both space and time. A case study involving electrical power data (in kw/h) weekly transmitted from photovoltaic plants is presented.

Chapter 3 describes applications of spatio-temporal interpolators in sensor networks. We describe two interpolation techniques, which use trend clusters to interpolate missing data. The former performs the estimation phase by using the Inverse Distance Weighting approach, while the latter uses Kriging. Both have been adapted to a sensor network scenario. We provide a detailed description of both techniques with experiments.

Chapter 4 discusses the problem of data surveillance in sensor networks. We describe a computation preserving technique, which employees an incremental learning strategy to continuously maintain trend clusters referring to the most recent past of the sensor network activity. The analysis of trend clusters permits the search for possible change in the data, as well the production of forecasts of the future.

The book concludes with an examination of some sensor data analysis applications. Chapter 5 illustrates a business intelligence solution to monitor the efficiency of the energy production of photovoltaic plants and a data mining solution for fault detection in photovoltaic plants.

## Remarks

The future will witness large deployments of sensor networks. These networks of small devices will change our lifestyle. With the advances in their data mining ability, these networks will play increasingly important roles in smart cities, by being integrated into smart houses, offices, and roads. The evolution of the smart city idea follows the same line as computation: first hardware, then software, then data, and orgware. In fact, the smart city is joining with data sensing and data mining to generate new models in our understanding of cities.

We like to think that this book is a small step toward this future evolution. It is devoted to the description of general intelligent services across networks and the presentation of specific applications of these services in monitoring the efficiency of photovoltaic power plants. Networks are treated as online systems, whose origins lie in the way we are able to sense what is happening. Data mining is used to process sensed data and solve problems like monitoring energy production of photovoltaic plants.

## References

1. C.C. Aggarwal, An introduction to sensor data analytics, ed. by C.C. Aggarwal, *Managing and Mining Sensor Data* (Springer-Verlag, New York, 2013), pp. 1–8
2. V. Cantoni, L. Lombardi, P. Lombardi, Challenges for Data Mining in Distributed Sensor Networks, in *Proceedings of the 18th International Conference on Pattern Recognition —Vol (1), ICPR '06*, (IEEE Computer Society, Washington, USA, 2006), pp. 1000–1007
3. J. Elson, D. Estrin, *Wireless Sensor Networks*, Chapter sensor networks: a bridge to the physical world (Kluwer Academic Publishers, Norwell, 2004), pp. 3–20
4. J. Gama, M. Gaber, *Learning from Data Streams: Processing Techniques in Sensor Networks* (Springer, New York, 2007)
5. A.P. Jayasumana, *Sensor Networks—Technologies, Protocols and Algorithms* (Springer, Netherlands, 2009)
6. T. Palpanas, Real-time data analytics in sensor networks, ed. by C.C. Aggarwal *Managing and Mining Sensor Data* (Springer-Verlag, 2013) pp. 173–210

# Acknowledgements

This work has been carried out in fulfillment of the research objectives of the project "EMP3: Efficiency Monitoring of Photovoltaic Power Plants", funded by the "Fondazione Cassa di Risparmio di Puglia". The authors wish to thank Lynn Rudd for her help in reading the manuscript and Pietro Guccione for his comments and discussions on the manuscript.

# Contents

# Chapter 1
# Sensor Networks and Data Streams: Basics

**Abstract** Recent advances in pervasive computing and sensor technologies have significantly influenced the field of geosciences, by changing the type of dynamic environmental phenomena that can be detected, monitored, and reacted to. Another important aspect is the real-time data delivery of novel platforms. In this chapter, we describe the specific characteristics of sensor data and sensor networks. Furthermore, we identify the most promising streaming models, which can be embedded in intelligent sensor platforms and used to mine real-time data for a variety of analytical insights.

## 1.1 Sensor Data: Challenges and Premises

The continued trend toward miniaturization and inexpensiveness of sensor nodes has paved the way for the explosive living ubiquity of geosensor networks (GSNs). They are made up of thousands, even millions, of untethered, small-form, battery-powered computing nodes with various sensing functions, which are distributed in a geographic area. They allow us to measure geographically and densely distributed data for several physical variables (e.g. atmospheric temperature, pressure, humidity, or energy efficiency of photovoltaic plants), by shifting the traditional centralized paradigm of monitoring a geographical area from the macro-scale to the micro-scale.

Geosensor networks serve as a bridge between the physical and digital worlds and enable us to monitor and study dynamic physical phenomena at granularity details that were never possible before [1]. While providing data with unparalleled temporal and spatial resolution, geosensor networks have pushed the frontiers of traditional GIS research into the realms of data mining. Higher level spatial and temporal modeling needs to be enforced in parallel, so that users can effectively utilize the potential.

The major challenge of a geosensor network is to combine the sensor nodes in computational infrastructures. These are able to produce globally meaningful information

A. Appice et al., *Data Mining Techniques in Sensor Networks*,
SpringerBriefs in Computer Science, DOI: 10.1007/978-1-4471-5454-9_1,
© The Author(s) 2014

from data obtained by individual sensor nodes and contribute to the synthesis and communication of geo-temporal intelligent information. The infrastructures should use appropriate primitives to account for both the spatial dimension of data, which determines the ground location of a sensor, and the temporal dimension of data, which determines the ground time of a reading. Both are information-bearing and play a crucial role in the synthesis of intelligence information.

The spatial dimension yields spatial correlation forms [2] that anyone seriously interested in processing spatial data should take into account [3]. Spatial autocorrelation is the correlation among values of a single attribute strictly due to their relatively close locations on a two-dimensional surface. Intuitively, it is a property of random variables taking values, at pairs of locations a certain distance apart, that are more similar (positive autocorrelation) or less similar (negative autocorrelation) than expected for pairs of observations at randomly selected locations [2]. Positive autocorrelation is the most common in geographical phenomena [4], which is justified by Tobler's first law of geography, according to which "everything is related to everything else, but near things are more related than distant things" [5]. This law suggests that by picturing the spatial variation of a geophysical variable, measured by a sensor network over the map, we can observe zones where the distribution of data is smoothly continuous, with boundaries possibly marked by sharp discontinuities.

The temporal dimension determines the time extent of the data. In a statistical view of the network, the simplest case occurs when measurements of a sensor can be ascribed to a stationary process, i.e., the statistical features do not evolve at all. By contrast, in a geophysical context the statistical features tend to change over time. This violates the assumption of identical data distribution across time: the distribution of a field is usually subjected to time drift. However, statistical changes occur in general in long timescales, so that the evolution of a time series is predictable by using time correlations in data. There are several cases where time-evolving data are subjected to trends with slow and fast variations, possible seasonality, and cyclical irregularities. For example, trend and seasonality are properties of genuine interest in climatology [6] for which sensors are frequently installed.

Seeking spatial- and temporal-aware information in a geosensor network will bring numerous computational challenges and opportunities [7, 8] for collection, storage, and processing. These challenges arise from both accuracy and scalability perspectives. In this book, the challenges have been explored for the tasks of summarization, interpolation, and surveillance.

## 1.2 Data Mining

Data mining is the process of automatically discovering useful information in large data repositories. The three most popular data mining techniques are predictive modeling, clustering analysis, and anomaly analysis.

1. In predictive modeling, the goal is to develop a predictive model, capable of predicting the value of a label (or target variable) as a function of explanatory variables. The model is mined from historical data, where the label of each sample is known. Once constructed, a predictive model is used to predict the unknown label of new samples.
2. In cluster analysis, the goal is to partition a data set into groups of closely related data in such a way that the observations belonging to the same group, or cluster, are similar to each other, while the observations belonging to different clusters are not. Clusters are often used to summarize data.
3. In anomaly analysis, also called outlier detection, the goal is to detect patterns in a given data set that do not conform to an established normal behavior. The patterns thus detected are called anomalies and are often translated into critical, actionable information in several application domains. Anomalies are also referred to as outliers, change, deviation, surprise, aberrant, peculiarity, intrusion, and so on.

Data mining is a step of knowledge discovery in databases, the so-called KDD process for converting data into useful knowledge [9]. The KDD process consists of a series of steps; the most relevant are:

1. Data pre-processing, which transforms collected data into an appropriate form for subsequent analysis;
2. Actual data mining, which transforms the prepared data into patterns or models (prediction models, clusters, anomalies);
3. Post-processing of data mining results, which assesses the validity and usefulness of the extracted patterns and models and presents interesting knowledge to the final users by using visual metaphors or integrating knowledge into decision support systems.

Today, data mining is a technology that blends data analysis methods with sophisticated techniques for processing large data volumes. It also represents an active research field, which aims to develop new data analysis methods for novel forms of data. One of the frontiers of data research today is represented by spatiotemporal data [10], that is, observations of events that occur in a given place at a certain time, such as the data arriving from sensor networks. Here, the challenge is particularly tough: data mining tools are needed to master the complex dynamics of sensors which are distributed over a (large) region, produce a continuous flow of data, eventually at high speeds, act in dynamic, time-changing environments, etc. These issues require the design of appropriate, efficient data mining techniques for processing spatiotemporal data produced by sensor networks.

## 1.3 Snapshot Data Model

Without loss of generality, the following four premises describe the geosensor scenario that we have considered for this study.

1. Sensors are labeled with a progressive number within the network and they are georeferenced by means of 2-D point coordinates (e.g., latitude and longitude).
2. Spatial location of the sensors is known, distinct, and invariant, while the number of sensors, which acquire data, may change in time: a sensor may be temporally inactive and not acquire any measure for a time interval.
3. Active sensors acquire a stream of data for each numeric physical variable and acquisition activity is synchronized on the sensors of the network.
4. Time points of the stream are equally spaced in time.

A snapshot model, originally presented in [11], can then be used to represent sensor data which are georeferenced and timestamped. Let us consider an equal-width discretization of a time line $T$ and a numeric physical variable $Z$ for which georeferenced values are sampled by a geosensor network $K$ at the consecutive time points of $T$.

**Definition 1.1 (Data snapshot)** A data snapshot timestamped at $t$ (with $t \in T$) is the pair:

$$\langle K_t, z_t() \rangle, \tag{1.1}$$

where:

1. $K_t$ ($K_t \subseteq K$) is the set of sensors, which measures a value for $Z$ at the time point $t$.
2. $z_t()$ is a field function [12]:

$$z_t : K_t \mapsto Z, \tag{1.2}$$

which assigns the sensor $u \in K_t$ to the value $z_t(u)$ measured for the variable $Z$ from the sensor $u$ at time point $t$.

Though finite, $K_t$ may vary with time $t$, since sensors which operate in a network can change with the time. They can pass from being switched-on to being switched-off (and vice versa) in the network. Similarly, $z_t()$ may vary with $t$.

The data snapshots, which are acquired from a geosensor network $K$, produce a geodata stream (see Fig. 1.1).

**Definition 1.2 (Geodata stream)** In a geodata stream $z(T, K)$ the input elements $\langle K_{t_1}, z_{t_1}(K_{t_1}) \rangle, \langle K_{t_2}, z_{t_2}(K_{t_2}) \rangle, \ldots, \langle K_{t_i}, z_{t_i}(K_{t_i}) \rangle, \ldots$ arrive sequentially from $K$, snapshot by snapshot, at the consecutive time points of $T$ to describe geographically distributed values of $Z$.

The model of a geodata stream is, in general, an insert-only stream model [13], since once a data snapshot is acquired, it cannot be changed. Insert-only geodata are

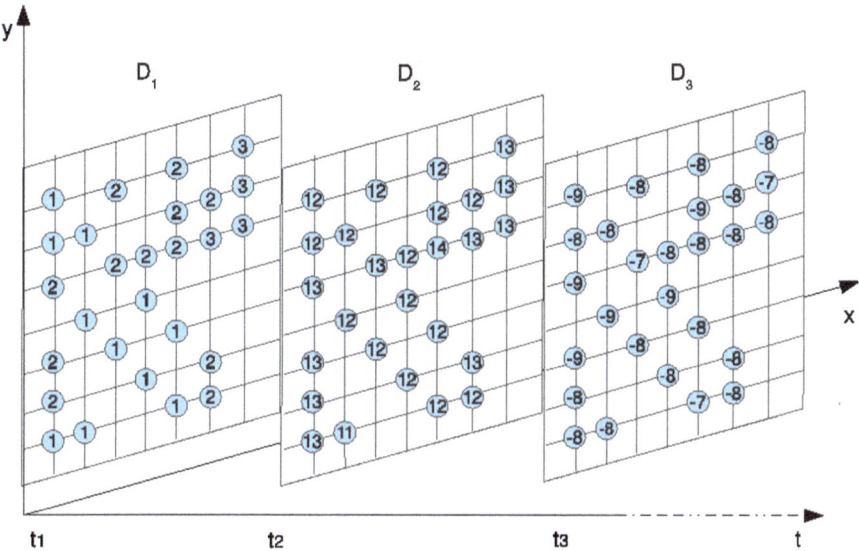

**Fig. 1.1** Snapshot representation of a geodata stream. A snapshot is timestamped with a discrete time point and snapshots continuously arrive at consecutive time points equally spaced in time. Sensors that are switched-on at a certain time are represented by blue circles in the snapshot. The number in a circle is the measure collected for a numeric physical variable $Z$ by the geosensor at the time point of the associated snapshot

collected in several environmental applications, such as determining trends in weather development [14] and pollution level of water [15] or tracking energy efficiency in sustainable energy systems [16].

## 1.4 Stream Data Model

Geodata streams, like any data stream, are unbounded in length. In addition, data collected with a geosensor network are geographically distributed. Therefore, they have not only a time dimension but also a space dimension. The amount of geographically distributed data acquired at a specific time point can be very large. Any future demand for analysis, which references past data, also becomes problematic. These are situations in which applying stream models to geodata become relevant.

It is impractical to store all the geodata of a stream. Looking for summaries of previously seen data is a valid alternative [17]. Summaries can be stored in place of the real data, which are discarded. This introduces a trade-off between the size of the summary and the ability to perform any future query by piecing together precise past data from summaries.

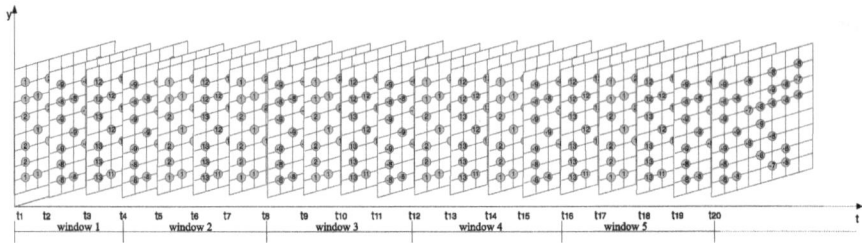

**Fig. 1.2** Count-based window model of a geodata stream with window size $w = 4$

Windows are commonly used stream approaches to query open-ended data. Instead of computing an answer over the whole data stream, the query (or operator) is computed, maybe several times, over a finite subset of snapshots. Several window models are defined in the literature. In the following subsections the most relevant ones are described.

## 1.4.1 Count-Based Window

A count-based window model [18] decomposes a stream into consecutive (non-overlapping) windows of fixed size (see Fig. 1.2). When a window is completed, it is queried. The answer is stored, while windowed data are discarded.

**Definition 1.3 (Count-based window model)** Let $w$ be the window size of the model. A count-based window model decomposes a geodata stream $z(T, K)$ in non-overlapping windows,

$$t_1 \stackrel{z(T,K)}{\to} t_w, \ t_{w+1} \stackrel{z(T,K)}{\to} t_{2w}, \ \ldots, \ t_{(i-1)w+1} \stackrel{z(T,K)}{\to} t_{iw}, \ \ldots \tag{1.3}$$

where the window $t_{(i-1)w+1} \stackrel{z(T,K)}{\to} t_{iw}$ is the series of $w$ data snapshots acquired at the consecutive time points of the time interval $[t_{(i-1)w+1}, t_{iw}]$ with $t_{(i-1)w+1}, t_{iw} \in T$.

## 1.4.2 Sliding Window

A sliding window model [18] is the simplest model to consider the recent data of the stream and run queries over the data of the recent past only. This type of window is similar to the first-in, first-out data structure. When a snapshot timestamped with $t_i$ is acquired and inserted in the window, another snapshot timestamped with $t_{i-w}$ is discarded (see Fig. 1.3), where $w$ represents the size of the window.

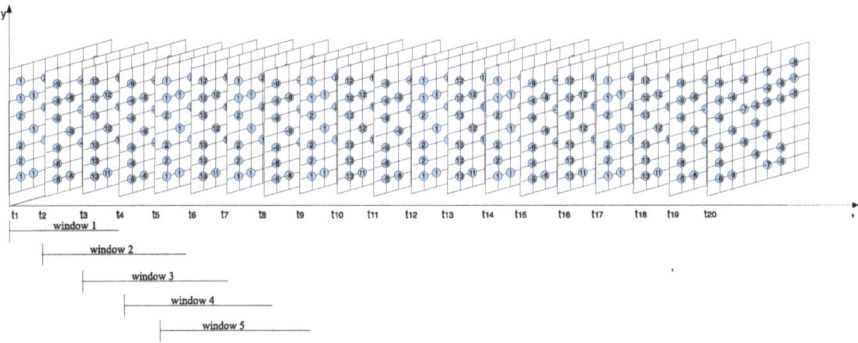

**Fig. 1.3** Sliding window model of a geodata stream with window size $w = 4$

**Definition 1.4 (Sliding window model)** Let $w$ be the window size of the model. A sliding window model decomposes the geodata stream $z(T, K)$ into overlapping windows,

$$t_1 \stackrel{z(T,K)}{\rightarrow} t_w, \ t_2 \stackrel{z(T,K)}{\rightarrow} t_{w+1}, \ \ldots, \ t_{i-w+1} \stackrel{z(T,K)}{\rightarrow} t_i, \ \ldots, \quad (1.4)$$

where the window $t_{i-w+1} \stackrel{z(T,K)}{\rightarrow} t_i$ is the series of $w$ data snapshots acquired at the consecutive time points of the time interval $[t_{i-w+1}, t_i]$ with $t_{i-w+1}, t_i \in T$.

The history for the snapshot $\langle K_{t_i}, z_{t_i}(k_{t_i}) \rangle$ is the window $t_{i-w} \stackrel{z(T,K)}{\rightarrow} t_{i-1}$.

## 1.5 Summary

The large deployments of sensor networks are changing our lifestyle. With these advances in computation power and wireless technology, networks start to play an important role in smart cities. Sensor networks consist of distributed autonomous devices that cooperatively monitor an environment. Each node in a sensor network is able to sense, process, and act. Data produced by sensor networks pose several issues: sensors are distributed; they produce a continuous stream of data, possibly at high speed; they act in dynamic time-changing environments; and the number of sensors can be very large and change with time and so on.

Mining data streams generated by sensor networks can play a central role in several applications, such as monitoring, tracking, and controlling. In this chapter, we provided a brief introduction to sensor data and sensor networks by focusing on challenges and opportunities for data mining. We revised basic models for data stream representation and processing.

# References

1. S. Nittel, Geosensor networks, in *Encyclopedia of GIS*, ed. by S. Shekhar, H. Xiong, (Springer, 2003)
2. P. Legendre, Spatial autocorrelation: trouble or new paradigm? Ecology **74**, 1659–1673 (1993)
3. J. LeSage, K. Pace, Spatial dependence in data mining, in *Data Mining for Scientific and Engineering Applications*, (Kluwer Academic Publishing, 2001), pp. 439–460
4. C. Sanjay, S. Shashi, W. Wu, Modeling spatial dependencies for mining geospatial data: An introduction, in *Geographic Data Mining and Knowledge Discovery*, (Taylor and Francis, 2001), pp. 131–159
5. W. Tobler, Cellular geography, in *Philosophy in Geography*, (1979), pp. 379–386
6. M. Mudelsee, in *Climate Time Series Analysis*, Atmospheric and Oceanographic Sciences Library, vol 42 (Springer, 2010)
7. A. P. Jayasumana. Sensor networks - technologies, protocols and algorithms, 2009.
8. C. C. Aggarwal. An introduction to sensor data analytics, in *Managing and Mining Sensor Data*, ed. by C. C. Aggarwal (Springer-Verlag, 2013), pp. 1–8
9. U. Fayyad, G. Piatesky-Shapiro, P. Smyth, R. Uthurusamy, *Advances in Knowledge Discovery and Data Mining*, (Mit Press, 1996)
10. M. Nanni, B. Kuijpers, C. Körner, M. May, D. Pedreschi, Spatiotemporal data mining, in *Mobility, Data Mining and Privacy: Geographic Knowledge Discovery*, ed. by F. Giannotti, D. Pedreschi, ( Springer-Verlag, 2008), pp. 267–296
11. C. Armenakis, Estimation and organization of spatio-temporal data, In *Proceedings of the Canadian Conference on GIS92*, 1992, p. 900-911
12. S. Shekhar, S. Chawla, *Spatial databases: A tour*, (Prentice Hall, 2003)
13. J. Gama, P. P. Rodriques, Data stream processing, in *Learning from Data streams: Processing Techniques in Sensor Networks*, ed. by J. Gama, M. M. Gaber (Springer, 2007)
14. D. Culler, D. Estrin, M. Srivastava, Guest editors' introduction: Overview of sensor networks. Computer **37**(8), 41–49 (2004)
15. A. Ostfeld, J. Uber, E. Salomons et al., The battle of the water sensor networks (BWSN): a design challenge for engineers and algorithms. J. Water Resour. Plan. Manage. **134**(6), 556 (2008)
16. Z. Zheng, Y. Chen, M. Huo, B. Zhao, An overview: the development of prediction technology of wind and photovoltaic power generation. Energy Procedia **12**, 601–608 (2011)
17. R. Chiky, G. Hébrail, Summarizing distributed data streams for storage in data warehouses, In *Proceedings of the 10th International Conference on Data Warehousing and Knowledge Discovery, DaWaK 2008*. LNCS, vol 5182, (Springer-Verlag, 2008), p. 65–74
18. M.M. Gaber, A. Zaslavsky, S. Krishnaswamy, Mining data streams: a review. ACM SIGMOD Rec. **34**(2), 18–26 (2005)

# Chapter 2
# Geodata Stream Summarization

**Abstract** The management of massive amounts of geodata collected by sensor networks creates several challenges, including the real-time application of summarization techniques, which should allow the storage of this unbounded volume of georeferenced and timestamped data in a server with a limited memory for any future query. SUMATRA is a summarization technique, which accounts for spatial and temporal information of sensor data to produce the appropriate trade-off between size and accuracy of geodata summarization. It uses the count-based model to process the stream. In particular, it segments the stream into windows, computes summaries window-by-window, and stores these summaries in a database. The trend clusters are discovered as a summary of each window. They are clusters of georeferenced data, which vary according to a similar trend along the time horizon of the window. Signal compression techniques are also considered to derive a compact representation of these trends for storage in the database. The empirical analysis of trend clusters contributes to assess the summarization capability, the accuracy, and the efficiency of the trend cluster-based summarization schema in real applications. Finally, a stream cube, called geo-trend stream cube, is defined. It uses trends to aggregate a numeric measure, which is streamed by a sensor network and is organized around space and time dimensions. Space-time roll-up and drill-down operators allow the exploration of trends from a coarse-grained and inner-grained hierarchical view.

## 2.1 Summarization in Stream Data Mining

The summarization task is well known in stream data mining, where several techniques, such as sampling, Fourier transform, histograms, sketches, wavelet transform, symbolic aggregate approximation (SAX), and clusters have been tailored to summarize data streams. The majority of these techniques were originally defined to summarize unidimensional and single-source data streams. The recent literature includes several extensions of these techniques, which address the task of summa-

A. Appice et al., *Data Mining Techniques in Sensor Networks*,
SpringerBriefs in Computer Science, DOI: 10.1007/978-1-4471-5454-9_2,
© The Author(s) 2014

rization in multidimensional data streams and, sometimes, multi-source data streams. A sensor network is a multi-source data stream generator.

### 2.1.1 Uniform Random Sampling

This is the easiest form of data summarization, which is suitable for summarizing both unidimensional and multidimensional data streams [1]. Data are randomly selected from the stream. In this way, summaries are generated fast, but the arbitrary dropping rate may cause high approximation error. Stratified sampling [2] is the alternative to uniform sampling to reduce errors, due to the variance in data.

### 2.1.2 Discrete Fourier Transform

This is a signal processing technique, which is adapted in [3] to summarize a stream of unidimensional numeric data. For each numeric value flowing in the stream, the Pearson correlation coefficient is computed over a stream window and the data, whose absolute correlation is greater than a threshold, are sampled. To the best of our knowledge, no other present work investigates the discrete Fourier transforms into multidimensional data streams and multi-source data streams.

### 2.1.3 Histograms

These are summary structures used to capture the distribution of values in a data set. Although histogram-based algorithms were originally used to summarize static data, several kinds of histograms have been proposed in the literature for the summarization of data streams. In Refs. [4, 5], V-Optimal histograms are employed to approximate the distribution of a set of values by a piecewise constant function, which minimizes the squared error sum. In Ref. [6], equiwidth histograms partition the domain into buckets, such that the number of values falling in a bucket is uniform across the buckets. Quantiles of the data distributions are maintained as bucket boundaries. End-biased histograms [7] maintain exact counts of items that occur with a frequency above a threshold and approximate the other counts by uniform distribution. Histograms to summarize multidimensional data streams are proposed in [8, 9].

### 2.1.4 Sketches

These are approximation algorithms for data streams that allow the estimation of frequency moments and aggregates over joins [10]. A sketch is constructed by taking an inner product of the data distribution with a vector of random values chosen from some distribution with a known expectation. The accuracy of estimation will depend on the contribution of the sketched data elements with respect to the rest of the streamed data. The size of the sketch depends on the memory available, hence the accuracy of the sketch-based summary can be boosted by increasing the size of the sketch. Sketching and sampling have been combined in [11]. An adaptive sketching technique to summarize multidimensional data streams is reported in [12].

### 2.1.5 Wavelets

These permit the projection of a sequence of data onto an orthogonal set of basis vectors. The projection wavelet coefficients have the property that the stream reconstructed from the top coefficients best approximates the original values in terms of the squared error sum. Two algorithms that maintain the top wavelet coefficients as the data distribution drifts in the stream are described in [10] and [13], respectively. Multidimensional Haar synopsis wavelets are described in [13].

### 2.1.6 Symbolic Aggregate Approximation

This is a symbolic representation, which allows the reduction of a numeric time series to a string of arbitrary length [14]. The time series is first transformed in the Piecewise Aggregate Approximation (PAA) and then the PAA representation is discretized into a discrete string. The important characteristic of this representation is that it allows a distance measure between symbolic strings which lower bounds the true distance between the original time series. Up to now, the utility of this representation has been investigated in clustering, classification, query by content, and anomaly detection in the context of motif discovery, but the data reduction it operates opens opportunities for the summarization task.

### 2.1.7 Cluster Analysis

Cluster analysis is a summarization paradigm which underlines the advantage of discovering summaries (clusters) that adjust well to the concept drift of data streams. The seminal work is that of Aggarwal et al. [15], where a k-means algorithm is tailored to discover micro-clusters from multidimensional transactions which arrive in a stream. Micro-clusters are adjusted each time a transaction arrives, in order to preserve the temporal locality of data along a time horizon. Clusters are compactly represented by means of cluster feature vectors, which contain the sum of timestamps along the time horizon, the number of clustered points and, for each data dimension, both the linear sum and the squared sum of the data values.

Another clustering algorithm to summarize data streams is presented in [16]. The main characteristic of this algorithm is that it allows us to summarize multi-source data streams. The multi-source stream is composed of sets of numeric values which are transmitted by a variable number of sources at consecutive time points. Timestamped values are modeled as 2D (time-domain) points of a Euclidean space. Hence, the source position is neither represented as a dimension of analysis nor processed as information-bearing. The stream is broken into windows. Dense regions of 2D points are detected in these windows and represented by means of cluster feature vectors. A wavelet transform is then employed to maintain a single approximate representation of cluster feature vectors, which are similar over consecutive windows. Although a spatial clustering algorithm is employed, the aim of taking into account the spatial correlation of data is left aside.

Ma et al. [17] propose a cluster-based algorithm, which summarizes sensor data headed by the spatial correlation of data. Sensors are clustered, snapshot by snapshot, based on both value similarity and spatial proximity of sensors. Snapshots are processed independently of each other, hence purely spatial clusters are discovered without any consideration of a time variant in data. A form of surveillance of the temporal correlation on each independent sensor is advocated in [18], where the clustering phase is triggered on the remote server station only when the status of the monitored data changes on sensing devices. Sensors keep online a local discretization of the measured values. Each discretized value triggers a cell of a grid by reflecting the current state of the data stream at the local site. Whenever a local site changes its state, it notifies the central server of its new state.

Finally, Kontaki et al. [19] define a clustering algorithm, which is out of the scope of summarization, but originally develops the idea of the trend to group time series (or streams). A smoothing process is applied to identify the time series vertexes, where the trend changes from up to down or vice versa. These vertexes are used to construct piecewise lines which approximate the time series. The time series are grouped in a cluster, according to the similarity between the associated piecewise lines. In the case of streams, both the piecewise lines and the clusters are computed incrementally in sliding windows of the stream. Although this work introduces the idea of a trend as the base for clustering, the authors neither account for the spatial distribution of a cluster, grouped around a trend, nor investigate the opportunity of a compact representation of these trends for the sake of summarization. This idea has inspired the trend cluster based summarization technique introduced in [20] and is described in the rest of this chapter.

## 2.2 Trend Cluster

A trend cluster is a spatiotemporal pattern, recently defined in [20], to model the prominent temporal trends in the positive spatial autocorrelation of a geophysical numerical variable monitored through a sensor network. It is a cluster of neighbor

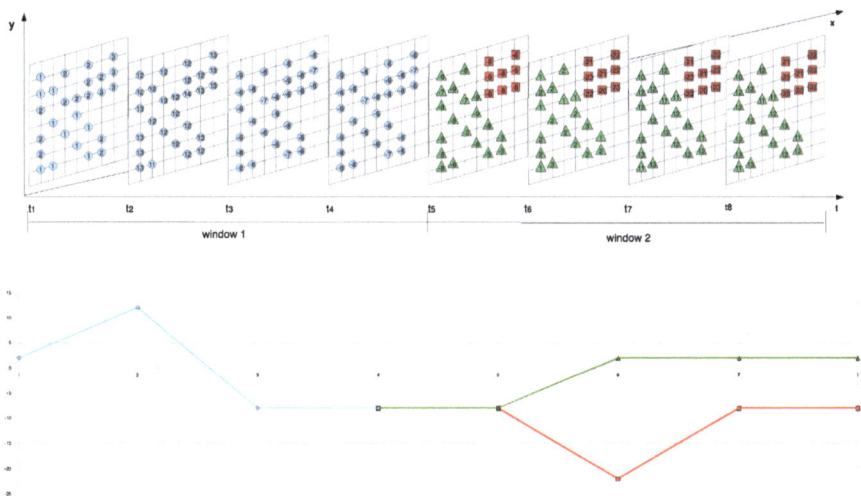

**Fig. 2.1** Trend clusters on a count-based model of the geodata stream ($w = 4$). The blue cluster groups circle sensors, whose values vary as the blue polyline from $t_1$ to $t_4$. The red cluster groups squared sensors, whose values vary as the red polyline from $t_5$ to $t_8$. The green cluster groups triangular sensors, whose values vary as the green (colored) polyline from $t_5$ to $t_8$

sensors, which measure data, whose temporal variation, called *trend polyline*, is similar over the *time horizon* of the window (see Fig. 2.1).

**Definition 2.1** (**Trend Cluster**) Let $z(T, K)$ be a geodata stream. A trend cluster is the triple:

$$(t_i \rightarrow t_j, \mathscr{C}, \mathscr{Z}), \qquad (2.1)$$

where:

1. $t_i \rightarrow t_j$ is a time horizon on $T$;
2. $\mathscr{C}$ is a set of "neighbor" sensors of $K$ measuring data for $Z$, which evolve with a "similar trend" from $t_i$ to $t_j$; and
3. $\mathscr{Z}$ is a time series representing the "trend" for data of $Z$ from $t_i$ to $t_j$. Each point in the time series can be a set of aggregating statistics (e.g., median or mean) of data for $Z$ measured by the sensors enumerated in $\mathscr{C}$.

In the count-based window model the time horizon is that of the count-based window, while in the sliding window model the time horizon is that of the sliding window.

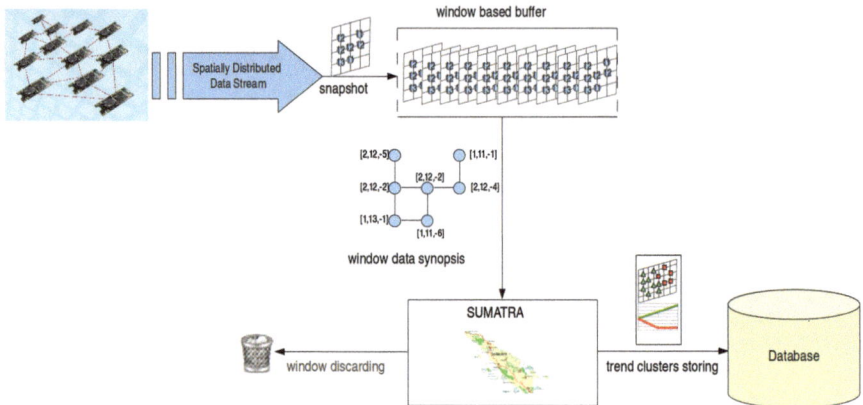

**Fig. 2.2** SUMATRA framework

## 2.3 Summarization by Trend Cluster Discovery

SUMATRA is a summarization algorithm, which resorts to the count-based stream model to process a geodata stream. It is now designed for the deployment on the powerful master nodes of a tiered sensor network.[1] It computes trend clusters along the time horizon of a window and derives a compact representation of the computed trends which is stored in a database (see Fig. 2.2). A buffer consumes snapshots as they arrive and pours them window-by-window into SUMATRA. The summarization process is three-stepped:

1. snapshots of a window are buffered into the data synopsis;
2. trend clusters are computed;
3. the window is discarded from the data synopsis, while trend clusters are stored in the database.

By using the count-based window, the time horizon is that of the window. It is implicitly defined by the enumerative code of the window when the window size $w$ is known. The storage of a trend cluster in a database (see Fig. 2.3) includes the window number, the identifiers of the sensors grouped into the cluster, and a representation of the trend polyline.

Input parameters for trend cluster discovery are the window size $w$ ($w > 1$), the neighborhood distance $d$, and a domain similarity threshold $\delta$. Input parameters for the trend polyline compression are either the error threshold $\varepsilon$ or the compression degree threshold $\sigma$. Both $\delta$ and $\varepsilon$ can influence the accuracy of the summary.

---

[1] The investigation of the in-network modality for this anomaly detection service is postponed to future developments of this study.

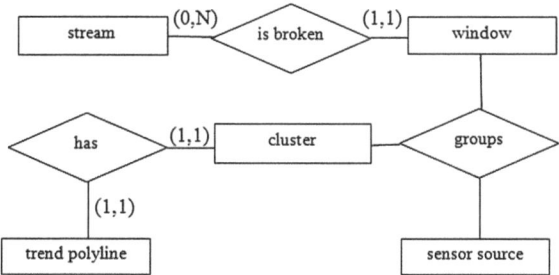

**Fig. 2.3**  Entity-relationship schema of the database where the trend clusters are stored

## 2.3.1 Data Synopsis

Snapshots of a window $W$ are buffered into a data synopsis $S$ which comprises a contiguity graph structure $G$ and a table structure $H$ (see Fig. 2.4).

Graph $G$ allows us to represent the discrete spatial structure, which is implicitly defined by the spatial location of sensors. It is composed of a node set $\mathcal{N}$ and an edge relation $\mathcal{E}$ with $\mathcal{E} \subseteq \mathcal{N} \times \mathcal{N}$. $\mathcal{N}$ is the set of active sensors which measure at least one value, for the variable $Z$, along the time horizon of $W$. Each node of $\mathcal{N}$ is labeled with the identifier of the associated sensor in the network. $\mathcal{E}$ is populated according to a user-defined distance relation (e.g., nearby within the radius $d$), which is derivable from the spatial location of each sensor [21]. In practice $(u, v) \in \mathcal{E}$ iff $distance(u, v) \leq d$. As the spatial locations of sensors are known and invariant, once the radius is set, the distance between each pair of sensors is always computable and does not change with time. Despite this fact, the structure of $G$ is subject to change at each new window $W$, which is completed in the stream: sensors may become active or inactive along the time horizon of a window, hence, associated nodes are added to or removed from the graph together with the connecting edges.

Table $H$ is a bidimensional matrix; rows correspond to the active sensors (or equivalently to the nodes of $\mathcal{N}$) and columns correspond to snapshots of the window. The $w$ measures collected for $Z$ from a node are stored in the tabular entries of the associated row of $H$. The one-to-one association between the graph nodes (keys) and the table rows (values) is made by means of a hash function. The collisions are managed according to traditional techniques designed for hash map data structure. In this chapter, the access to each value within the table row is abstractly denoted by means of the column index ranging between 1 and $w$. Thus, $H[u][t]$ denotes the tabular entry, which stores the value measured from the node $u$ at the $t$th snapshot of the window $W$. Missing values can be stored in $H$ in the presence of sensors which measure a value at one or more snapshots of the window, but they do not perform the measurement at all the snapshots of the window. They are preprocessed on-the-fly and replaced by an aggregate (median) of values stored in the corresponding row of the table.

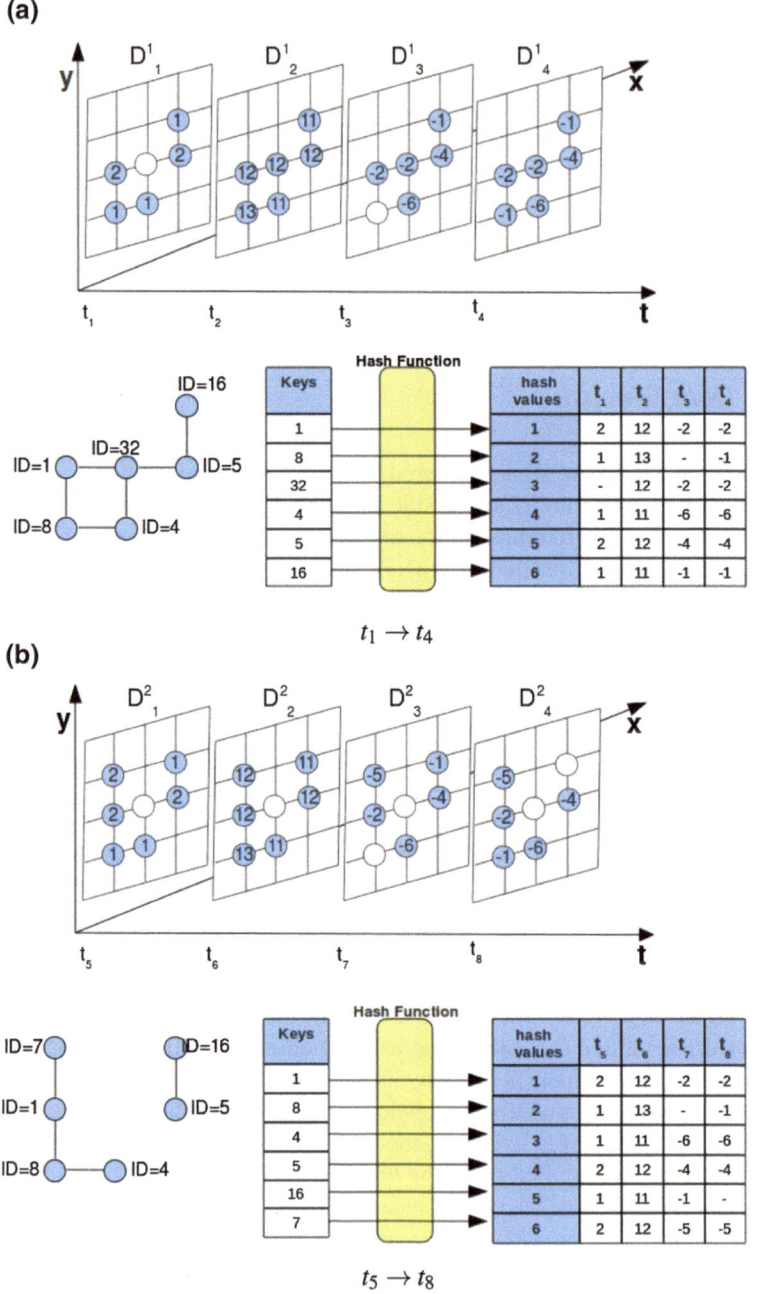

**Fig. 2.4** Four ($w = 4$) consecutive snapshots (windows) are stored in the data synopsis $S$. **a** $t_1 \to t_4$.
**b** $t_5 \to t_8$

### 2.3.2 Trend Cluster Discovery

We introduce definitions, which are preparatory to the presentation of the trend cluster discovery, then we illustrate the trend cluster discovery algorithm.

#### 2.3.2.1 Basic Concepts and Definitions

First, we define the relation of $\mathscr{E}$-reachability within a node set.

**Definition 2.2** ($\mathscr{E}$-**reachability relation**) Let $\mathscr{C}$ be a subset of $\mathscr{N}$ ($\mathscr{C} \subseteq \mathscr{N}$) and $u$ and $v$ be two nodes of $\mathscr{C}$ ($u, v \in \mathscr{C}$). $u$ is $\mathscr{E}$-reachable from $v$ in $\mathscr{C}$ iff:

1. $\langle u, v \rangle \in \mathscr{E}$ (direct $\mathscr{E}$-reachability, i.e., distance $(u, v) \leq d$), or
2. $\exists r \in \mathscr{C}$, such that $\langle u, r \rangle \in \mathscr{E}$ and $r$ is $\mathscr{E}$-reachable from $v$ in $\mathscr{C}$ (transitive $\mathscr{E}$-reachability).

Then, we define the property of $\mathscr{E}$-feasibility of a node set.

**Definition 2.3** ($\mathscr{E}$-**feasibility**) Let $\mathscr{C}$ be a subset of $\mathscr{N}$. $\mathscr{C}$ is feasible with the relation $\mathscr{E}$ iff:

$$\forall p, q \in \mathscr{C}: p \text{ is } \mathscr{E}\text{-reachable from } q \text{ in } \mathscr{C} \quad \text{(or vice versa).} \tag{2.2}$$

The trend polyline prototype associated with a node set is defined below.

**Definition 2.4** (**Trend polyline prototype**) Let $\mathscr{C}$ be a subset of $\mathscr{N}$. The trend polyline prototype of $\mathscr{C}$, denoted by $\mathscr{Z}$, is the chain of straight-line segments connecting the $w$ vertexes of the time series, which is defined as follows:

$$\mathscr{Z} = [(1, \mathscr{Z}(1)), (2, \mathscr{Z}(2)), \dots, (w, \mathscr{Z}(w))], \tag{2.3}$$

where $\mathscr{Z}(t)$ ($t = 1, 2, \dots, w$) is the aggregate (e.g. median) of values measured by nodes of $\mathscr{C}$ at the $t$th snapshot of the window (i.e. $\mathscr{Z}(t) = \text{aggregate}(\{H[u][t] \mid u \in \mathscr{C}\})$).

Finally, we define the property of the trend purity of a node set.

**Definition 2.5** ($\delta$-**bounded trend purity**) Let

1. $\delta$ be a user-defined domain similarity threshold;
2. $\mathscr{C}$ be a subset of $\mathscr{N}$;
3. $\mathscr{Z}$ be the trend polyline prototype of $\mathscr{C}$.

The trend purity of $[\mathscr{C}, \mathscr{Z}]$ is a binary property defined as follows:

$$\text{purity}([\mathscr{C}, \mathscr{Z}]) = \begin{cases} \text{true} & \text{iff } \dfrac{\left( \sum\limits_{u \in \mathscr{C}} \text{sim}(u, \mathscr{Z}) \right)}{|\mathscr{C}|} = 1, \\ \text{false} & \text{otherwise} \end{cases} \tag{2.4}$$

where $|\mathscr{C}|$ is the cardinality of $\mathscr{C}$ and:

$$\text{sim}(u, \mathscr{Z}) = \begin{cases} 1 \text{ iff } \forall t = 1, \ldots, w : \|(H[u][t]) - \mathscr{Z}(t))\| \leq \delta \\ 0 \text{ otherwise.} \end{cases} \tag{2.5}$$

This similarity computation schema requires the computation of the distance between each sensed value and the centroid of the cluster.

Finally, we describe the form of the trend cluster, which will be computed by SUMATRA.

**Definition 2.6   (Trend cluster)** Based upon Definition 2.1, a trend cluster is the triple $(i, \mathscr{C}, \mathscr{Z})$, such that:

1. $i$ enumerates the window where the trend cluster is discovered;
2. $\mathscr{C}$ is a subset of $\mathscr{N}$ which is feasible with the relation $\mathscr{E}$ (see Definition 2.3);
3. $\mathscr{Z}$ is the trend polyline prototype of $\mathscr{C}$ (see Definition 2.4);
4. $[\mathscr{C}, \mathscr{Z}]$ satisfies the trend purity property (see Definition 2.5).

Based on Definition 2.6, we observe that a trend cluster corresponds to a completely connected subgraph of $G$, which exhibits a similar polyline evolution for data measured along the window time horizon (trend purity). The trend of the cluster is the polyline prototype according to which the trend cluster purity is evaluated. Then, intuitively, trend clusters can be computed by a graph-partitioning algorithm, which identifies subgraphs that are completely connected by means of the *strong* edges defined as follows.

**Definition 2.7   (Strong edge)** Let $\langle u, v \rangle$ be an edge of $\mathscr{E}$, then $\langle u, v \rangle$ is labeled as a *strong* edge in $\mathscr{E}$ iff, for each snapshot of the window $W$, the values measured from $u$ and $v$ differ from $\delta$ at worst, that is,

$$\forall t = 1, \ldots, w \colon \|H[u][t] - H[v][t]\| \leq \delta). \tag{2.6}$$

Informally, a strong edge connects nodes which exhibit a similar trend polyline evolution along the window time horizon. The strong edges are the basis for the computation of the strong neighborhood of a node.

**Definition 2.8   (Strong neighborhood)** Let $u$ be a node of $\mathscr{N}$, then the strong neighborhood of $u$, denoted by $\eta(u)$, is the set of nodes of $\mathscr{N}$ which are directly reachable from $u$ by means of strong edges of $\mathscr{E}$, that is,

$$\eta(u) = \{v | \langle u, v \rangle \in \mathscr{E} \text{ and } \langle u, v \rangle \text{ is strong}\}). \tag{2.7}$$

Based on Definition 2.8, a strong neighborhood, which can be seen as a set of nodes around a seed node, is feasible with respect to the edge relation and groups trend polylines with a similar evolution as the trend polyline of the neighborhood seed. These considerations motivate our idea of constructing trend clusters by merging

**(a)**                                         **(b)**

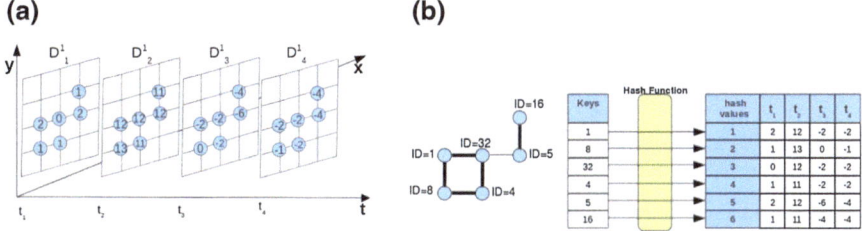

**Fig. 2.5** An example of window storage in SUMATRA. **a** A window of snapshots. **b** Window storage in the data synopsis

overlapping strong neighborhoods provided the resulting cluster satisfies the trend purity property.

### 2.3.2.2   The Algorithm

The top-level description of the trend cluster discovery is reported in Algorithm 2.1.

The discovery process is triggered each time a new window (Fig. 2.5a) is buffered into the data synopsis (Fig. 2.5b).

The computation starts by assigning $k = 1$, where $k$ enumerates the computed trend clusters. An unclustered node $u$ is randomly chosen as the seed of a new empty cluster $\mathscr{C}_k$. Then $u$ is added to $\mathscr{C}_k$ (the green cluster in Fig. 2.6a) and the trend polyline prototype $\mathscr{Z}_k$ is constructed (by calling *polylinePrototype*($\cdot$)). Both $\mathscr{C}_k$ and $\mathscr{Z}_k$ are expanded by using $u$ as the seed of the expansion process (by calling *expandCluster*($\cdot$, $\cdot$, $\cdot$)). The expanded trend cluster $[i, \mathscr{C}_k, \mathscr{Z}_k]$ is added to the pattern set $P$. $k$ is incremented by one and the clustering process is iteratively repeated until all nodes are assigned to a cluster (Fig. 2.6e, f).

The expansion process is described in Algorithm 2.2. The expansion of $[\mathscr{C}_k, \mathscr{Z}_k]$ is driven by a seed node $u$ and it is recursively defined. First, the strong neighborhood $\eta(u)$ is constructed by considering the unclustered nodes (by calling *neighborhood*($\cdot$, $\cdot$)). Then, the candidate cluster $\mathscr{C}' = \mathscr{C}_k \cup \eta(u)$ and the associated trend polyline prototype $\mathscr{Z}'$ are computed. The trend purity of $[\mathscr{C}', \mathscr{Z}']$ is computed (by calling *polylinePurity*($\cdot$, $\cdot$)). Two cases are distinguished:

1. $[\mathscr{C}', \mathscr{Z}']$ satisfies the trend purity property, then nodes of $\eta(u)$ are clustered into $\mathscr{C}_k$ (the green cluster in Fig. 2.6b) and the last computed $\mathscr{Z}'$ is assigned to $\mathscr{Z}_k$.
2. $[\mathscr{C}', \mathscr{Z}']$ does not satisfy the trend purity property and the addition of each node of $\eta(u)$ to $\mathscr{C}_k$ is evaluated node-by-node.

In both cases, nodes newly clustered in $\mathscr{C}_k$ are iteratively chosen as seeds to continue the expansion process (the gray circle in Fig. 2.6c). The expansion process stops if no new node is added to the cluster (the green cluster in Fig. 2.6d).

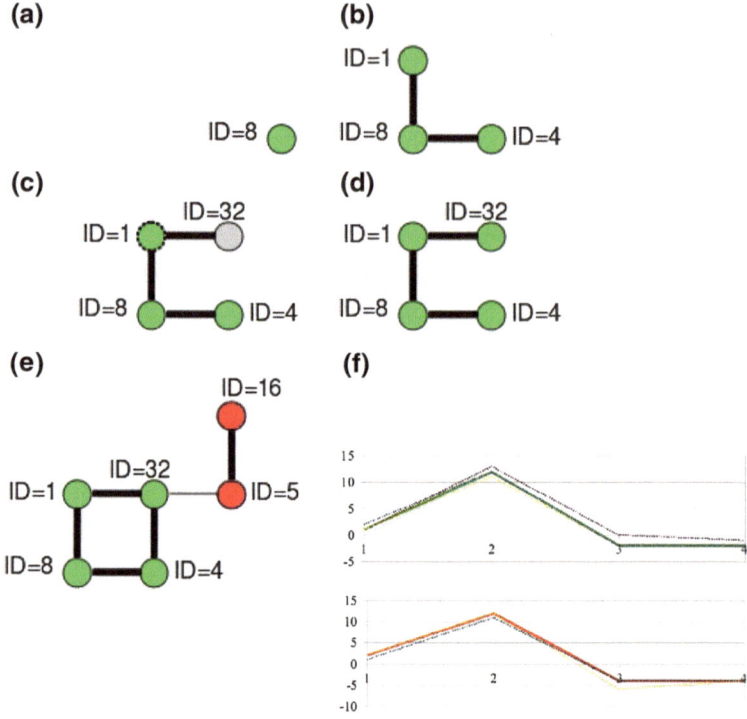

**Fig. 2.6** An example of trend cluster discovery in SUMATRA. **a** Cluster seed selection. **b** Strong neighborhood. **c** Expansion seed. **d** Complete cluster. **e** Clusters. **f** Trend polylines

### 2.3.2.3 Time Complexity

The time complexity of the trend cluster discovery is mostly governed by the number of neighborhood() invocations. At worst, one neighborhood is computed for each

---

**Algorithm 2.1** TrendClusterDiscovery($i$, $S[G, H]$, $\delta$) $\mapsto$ $P$

---

**Require:** $i$: the number which enumerates $W$
**Require:** $S[G, H]$: an instance of data synopsis $S$, where snapshots of $W$ are loaded
**Require:** $\delta$: the domain similarity threshold
**Ensure:** $P$: the set of trend clusters $[i, \mathscr{C}_k, \mathscr{Z}_k]$ discovered in $W$
1: $k \leftarrow 1$
2: **for all** $u \in \mathscr{N}$ **do**
3:   **if** $u$ is *UNCLUSTERED* **then**
4:     $[\mathscr{C}_k, \mathscr{Z}_k] \leftarrow$ expandCluster($\{u\}$, polylinePrototype($\{u\}$), $u$)
5:     append($P$, $[i, \mathscr{C}_k, \mathscr{Z}_k]$)
6:     $k \leftarrow k + 1$
7:   **end if**
8: **end for**

---

---

**Algorithm 2.2** expandCluster $(\mathscr{C}_k, \mathscr{L}_k, u) \mapsto [\mathscr{C}_k, \mathscr{L}_k]$

---

**Require:** $\mathscr{C}_k$: the node cluster
**Require:** $\mathscr{L}_k$: the trend polyline prototype of $\mathscr{C}_k$
**Require:** $u$: the seed node for the cluster expansion
**Ensure:** $[\mathscr{C}_k, \mathscr{L}_k]$: the expanded trend cluster
1: $\eta(u) \leftarrow$ neighborhood$(u)$
2: $[\mathscr{C}', \mathscr{L}'] \leftarrow [\mathscr{C}_k \cup \eta(u), \text{polylinePrototype}(\mathscr{C}_k \cup \eta(u))]$
3: **if** polylinePurity$(\mathscr{C}', \mathscr{L}')$ **then**
4:    $[\mathscr{C}_k, \mathscr{L}_k] \leftarrow [\mathscr{C}', \mathscr{L}']$
5:    **for all** $v \in \eta(u)$ **do**
6:        $[\mathscr{C}_k, \mathscr{L}_k] \leftarrow$ expandCluster$(\mathscr{C}_k, \mathscr{L}_k, v)$
7:    **end for**
8: **else**
9:    **for all** $v \in \eta(u)$ **do**
10:       $[\mathscr{C}', \mathscr{L}'] \leftarrow [\mathscr{C}_k \cup v, \text{polylinePrototype}(\mathscr{C}_k \cup v)]$
11:       **if** polylinePurity$(\mathscr{C}', \mathscr{L}')$ **then**
12:           $[\mathscr{C}_k, \mathscr{L}_k] \leftarrow$ expandCluster$(\mathscr{C}', \mathscr{L}', v)$
13:       **end if**
14:    **end for**
15: **end if**

---

sensor and evaluated in space and time. By using an indexing structure to execute such a neighborhood query and a quickselect algorithm (having linear time complexity) to compute the median aggregate, the time complexity of the trend cluster discovery in a window of $k$ nodes and $w$ snapshots is, at worst,

$$O(k(\underbrace{wlogk}_{neighbourhood()} + \underbrace{kw}_{polylinePrototype()} + \underbrace{kw}_{polylinePurity()})).$$

## 2.3.3 *Trend Polyline Compression*

A trend polyline is a time series that can be compressed by using any signal compression technique. We have investigated both Discrete Fourier Transform and Haar Wavelet. Both techniques take a trend cluster polyline $\mathscr{L}$ as input, transform $\mathscr{L}$ into $\mathscr{L}'$ and return $\mathscr{L}'$ as output for the storage in the database $DB$ [22]. Details of these techniques, including the inverse transforms and the strategies to control the compression degree or the compression error, are described in the following subsections.

### 2.3.3.1 Discrete Fourier Transform

The Discrete Fourier Transform (DFT) [23] is a technique of the Fourier analysis, which allows us to decompose $\mathscr{L}$ into a linear combination of orthogonal complex sinusoids, differing from each other in frequency. The coefficients of the linear combination represent $\mathscr{L}$ in the frequency domain of the sinusoidal basis. The DFT representation of $\mathscr{L}$ is then used to compute $\mathscr{L}'$.

Let $\mathscr{L}(1), \mathscr{L}(2), \ldots, \mathscr{L}(w)$ be the series of the $w$ values of $\mathscr{L}$ as they are equally spaced in time. The DFT permits us to define each $\mathscr{L}(t)$ as an instance of the linear combination of $w$ complex sinusoidal functions, as follows:

$$\mathscr{L}(t) = \frac{1}{w} \sum_{h=0}^{w-1} Z_h e^{\iota 2\pi \frac{h}{w}(t-1)}, \text{ with } t = 1, 2, \ldots, w, \qquad (2.8)$$

where $\iota$ is the imaginary unit and $e^{-\iota \frac{(2\pi)}{w} h(t-1)}$ represents the complex sinusoid with length $w$ and discrete frequency $h/w$. We observe that the frequency of the complex sinusoidal basis in Eq. (2.8) ranges between zero and $1/2$ (the so-called Nyquist frequency), as each complex sinusoid with $h/w$ greater than $1/2$ is equivalent to the complex sinusoid with frequency $(w-h)/w$ and the opposite phase.

The complex coefficients $Z_h$ are computed as follows:

$$Z_h = \sum_{t=1}^{w} \mathscr{L}(t) e^{-\iota\, 2\pi \frac{h}{w}(t-1)}, \text{ with } h = 0, 1, \ldots, w-1. \qquad (2.9)$$

Considering that coefficients $Z_h$ satisfy the Hermitian symmetry property,[2] it is sufficient to compute coefficients $Z_h$, with $h$ ranging between 0 and $w/2$. Other coefficients are achieved by the Hermitian symmetry property.

$\mathscr{L}'$ is computed by selecting the top $k$ coefficients $Z_h$ (with $k \leq w/2 + 1$). This coefficient selection is motivated by considerations reported in [23], which are well founded if $\mathscr{L}$ is a slow-time varying polyline. For this kind of polyline, central coefficients (i.e. the closest to the Nyquist coefficient) capture the short-term fluctuations of $\mathscr{L}(1)$, then they can be neglected with a minimal loss of information. This process is called low-pass filtering [23].

The inverse transform $\tau_{\mathscr{L}'} : \{1, 2, \ldots, w\} \mapsto \mathbb{R}$ to construct $\hat{\mathscr{L}}$ from $\mathscr{L}'$ is computed as follows:

$$\tau_{\mathscr{L}'}(t) = \frac{1}{w} \sum_{h=0}^{w-1} \tilde{Z}_h e^{\iota\, 2\pi \frac{h}{w}(t-1)} \text{ with } \tilde{Z}_h = \begin{cases} Z_h & h = 0, \ldots, k-1 \\ 0 & h = k, \ldots, w-k-1 \\ \overline{Z_{w+1-h}} & h = w-k, \ldots, w-1 \end{cases},$$
$$(2.10)$$

where $^{-}$ denotes the complex coniugate.

___

[2] $Z_h$ and $Z_{w-h}$ are complex conjugates [23].

### 2.3.3.2 Discrete Haar Wavelet

The Discrete Haar Wavelet (DHW) [24] is a kind of wavelet, which decomposes $\mathcal{L}$ into the linear combination of orthogonal functions, which are localized in time and represent short local subsections of the polyline.

The DHW defines each $\mathcal{L}(t)$ of the trend polyline $\mathcal{L}$ by means of the linear combination of the father function, the mother function, and the child functions, that is,

$$\mathcal{L}(t) = \alpha\phi(t-1) + \sum_{h=1}^{w-1} \beta_h \psi_h(t-1) \text{ with } t = 1, 2, \ldots, w, \tag{2.11}$$

where the father $\phi(\cdot)$ and the mother $\psi(\cdot)$ are defined as follows:

$$\phi(t) = 1 \qquad\qquad t = 0, \ldots, w-1 \tag{2.12}$$

$$\psi(t) = \psi_0(t) = \begin{cases} -1 & 0 \le t < w/2 \\ 1 & w/2 \le t < w \end{cases}, \tag{2.13}$$

while the $w-1$ child functions $\psi_h$ are defined as follows:

$$\psi_h(t) = 2^{\frac{n}{2}} \psi(2^n t - 1) \quad t = 0, \ldots, w-1, \tag{2.14}$$

with $n = \lfloor \log_2(h) \rfloor$ and $l = h \mod 2^n$. Each child $\psi_h$ has the shape of the mother $\psi$, but it is rescaled by a factor of $2^{n/2}$ and shifted by a factor of $l$. The coefficients $\alpha$ and $\beta_h$ (with $h = 1, 2, \ldots, w-1$) are computed as follows:

$$\alpha = \sum_{j=1}^{w} \mathcal{L}(t)\phi(t-1) \text{ and } \beta_h = \sum_{t=1}^{w} \mathcal{L}(t)\psi_h(t-1) \text{ with } h = 1, 2, \ldots, w-1. \tag{2.15}$$

As a filtering technique to compute $\mathcal{L}'$, the $k$ coefficients which are the largest in absolute value are retained. Thus, the root mean squared error between $\mathcal{L}$ and the polyline, reconstructed from $\mathcal{L}'$, is minimized [25]. The Haar Wavelet filtering technique does not retain coefficients $\beta_h$ as they are ordered according to $h$.

The inverse transform $\tau_{\mathcal{L}'} : \{1, 2, \ldots, w\} \mapsto \mathbb{R}$ to construct $\hat{\mathcal{L}}$ from $\mathcal{L}'$ is computed as follows:

$$\tau_{\mathcal{L}'}(t) = (\tilde{\alpha} + \sum_{h=1}^{w-1} \tilde{\beta}_h \psi_h(t-1) \text{ with } \tilde{\alpha}(\tilde{\beta}_h) = \begin{cases} \alpha(\beta_h) & \text{if } \alpha(\beta_h) \in \mathcal{L}' \\ 0 & \text{otherwise} \end{cases}. \tag{2.16}$$

### 2.3.3.3 Polyline Compression Analysis

First we make some considerations on the amount of information (number of bytes) necessary to store $\mathscr{L}'$ in the database. Then, we state the conditions under which we guarantee that $\mathscr{L}'$ is a compact representation of $\mathscr{L}'$.

**Proposition 2.1** *Let $\mathscr{L}$ be a trend polyline having size w, then the size of $\mathscr{L}$ is $\sigma_F w$ (bytes), where $\sigma_F$ is the size of a real number.*

*Proof* The proposition can be proved when points of $\mathscr{L}$ are equally spaced in time. Then $\mathscr{L}$ can be stored as the series of $w$ float values $\mathscr{L}(t)$, without losing information.                                                                                   □

**Proposition 2.2** *Let $\mathscr{L}'$ be computed from $\mathscr{L}$ by DFT or DHW.*

$$size(\mathscr{L}') = \begin{cases} (\sigma_F + \sigma_F)k \ bytes & [DFT] \\ (\sigma_I + \sigma_F)k \ bytes & [DHW] \end{cases}, \tag{2.17}$$

*where k represents the number of transformed coefficients, $\sigma_F$ is the size of a real number, and $\sigma_I$ is the size of an integer number.*

*Proof* The proposition is proved by considering that a complex DFT coefficient is represented by real unity and imaginary unity (both float values). A DHW coefficient is a float value with an integer index.                                                      □

**Proposition 2.3** *$\mathscr{L}'$ is a compact representation of $\mathscr{L}$ (i.e., $size(\mathscr{L}') \leq size(\mathscr{L})$) if and only if $k \leq \kappa$ with:* $\kappa = \begin{cases} \frac{\sigma_F}{\sigma_F + \sigma_I} w & [DHW] \\ \frac{w}{2} & [DFT] \end{cases}$.

*Proof* This proposition is derived from Propositions 2.1 and 2.2.                      □

### 2.3.3.4 Tuning $k$

The size of $\mathscr{L}'$ linearly depends on $k$, which should be a user-defined parameter. The choice of $k$ can be automatically made by fixing a boundary for either the error of the inverse transform or the size of the signal compression.

Error-Based Tuning

Let $\hat{\mathscr{L}}$ be the trend polyline reconstructed from $\mathscr{L}'$ with the inverse transform $\tau_{\mathscr{L}'}$ and $\varepsilon$ be the user-defined upper bound threshold for the error of reconstruction. $e(\mathscr{L}, \hat{\mathscr{L}})$ is the root mean squared error of approximating $\mathscr{L}$ by $\hat{\mathscr{L}}$, that is,

$$e(\mathscr{Z}, \hat{\mathscr{Z}}) = \sqrt{\frac{1}{w} \sum_{t=1}^{w} \left(\mathscr{Z}(t) - \hat{\mathscr{Z}}(t)\right)^2}.$$ (2.18)

Then $k$ is chosen as

$$k = \min(\kappa, \ \kappa_\varepsilon),$$ (2.19)

where $\kappa$ is the maximum number of coefficients admitted in $\mathscr{Z}'$ (see Proposition 2.3) and $\kappa_\varepsilon$ is the minimum $k$ to guarantee a root mean squared error less than or equal to $\varepsilon$.

Formally,

$$\kappa_\varepsilon = \arg\min_k \{k \mid e(\mathscr{Z}, \hat{\mathscr{Z}}_k) \le \varepsilon\},$$ (2.20)

with $\hat{\mathscr{Z}}_k$ as the polyline reconstructed from $\mathscr{Z}'_k$ and $\mathscr{Z}'_k$ as the compact representation of $\mathscr{Z}$ which contains $k$ coefficients.

To determine $\kappa_\varepsilon$, the root mean squared error is computed in the transformed domain according to the Parseval identity. This identity states that the sum of the squared values in a domain is equal to the same sum computed in the transformed domain.[3] According to this identity, in the case of DFT, the root mean squared error is computed as the root mean of the squared filtered coefficients, that is,

$$e(\mathscr{Z}, \hat{\mathscr{Z}}_{k_\varepsilon}) = \sqrt{\frac{1}{w} \sum_{t=1}^{w} \left(\mathscr{Z}(t) - \hat{\mathscr{Z}}(t)\right)^2} \underbrace{=}_{Parseval\ identity} \sqrt{\frac{1}{w} \sum_{t=k}^{w-k} |\mathscr{Z}(t)|^2}.$$ (2.21)

The advantage of the Parseval identity is that it allows us to avoid the computation of $\hat{\mathscr{Z}}_{k_\varepsilon}$ to look for $k$. Considering that the coefficients of the transform are ordered in some way and that the filtering drops down the last coefficients of this order, we iteratively compute the sum of the squares of the coefficients, which are in the last positions of the ordering, until this sum approximates $\varepsilon$. Thus, $k_\varepsilon$ corresponds to the number of coefficients which are not summed to compute $\varepsilon$. Similarly, in the case of DHW, the Parseval identity allows us to compute the error in the domain of the Haar wavelet coefficients. Haar coefficients are ordered by the descending absolute value and, as for DFT, the filtering drops down the last coefficients.

### Size-Based Tuning

Let $\sigma$ be the user-defined upper bound for the degree of compression that $\mathscr{Z}'$ must produce with respect to $\mathscr{Z}$ (i.e. $\sigma \approx \frac{\text{size}(\mathscr{Z}')}{\text{size}(\mathscr{Z})}$). Then $k$ is computed as $k = \min(\kappa, \kappa_\sigma)$, such that, on the basis of Proposition 2.2, we have:

---

[3] This identity expresses in some way the law of conservation of energy.

$$k_\sigma = \begin{cases} \frac{\sigma}{(\sigma_I + \sigma_F)} w \ [\text{DHW}] \\ \frac{\sigma}{(\sigma_F + \sigma_F)} w \ [\text{DFT}] \end{cases} . \qquad (2.22)$$

#### 2.3.3.5  General Considerations

DFT and DHW may have different performances depending on the specific characteristics of the data. Fourier analysis, which decomposes a signal in a linear combination of periodic, regular functions (sinusoids), has particular characteristics to fit data generated according to periodic or close-to-periodic functions. The smoother the trend polyline (i.e., short-time fluctuations of data are negligible), the lower the number of coefficients required to achieve an accurate summary. Wavelet analysis, like Fourier transform, involves the representation of signals in terms of simpler functions, but the base functions of wavelets are fixed building blocks at different scales and positions. Hence, the Wavelet method, which has the discontinuity in the base, is more appropriate than the Fourier method for fitting non-periodic, wide-band signals with abrupt discontinuities. The comparative study, which inspires this analysis, is reported in [26].

## 2.4  Empirical Evaluation

SUMATRA, whose implementation is available to the public,[4] is written in Java and interfaces a database managed by a MySQL DBMS. The trend cluster discovery is evaluated on several real-world streams.

In the next subsection, we describe the geodata streams employed in this experimental study and describe the experimental setting. Subsequently, we present and comment on empirical results obtained with the geodata in this study.

### 2.4.1  Streams and Experimental Setup

We consider geodata streams, derived from both indoor and outdoor sensor networks, and evaluate the summarization performance in terms of accuracy and size of the summary, as well as the computation time spent summarizing the data. The experiments are performed on an Intel(R) Core(TM) 2 DUO CPU E4500 @2.20 GHz with 2.0 GiB of RAM Memory, running Ubuntu Release 11.10.

---

[4] http://www.di.uniba.it/~kdde/index.php/SUMATRA

### 2.4.1.1  Data Streams

The *Intel Berkeley Lab* (IBL) geodata stream[5] collects *indoor temperature* (in Celsius degrees) and *humidity* (in RH) measurements transmitted every 31 s from 54 sensors deployed in the Intel Berkeley Research lab, between February 28th and April 5th 2004. A sensor is considered spatially close to every other sensor in the range of six meters. The transmitted values are discontinuous and very noisy. Missing values occur in most snapshots, so the number of transmitting sensors is variable in time. By using a box plot, we deduce that air temperature values presumably range between 9.75 and 34.6, while the humidity values presumably range between 0 and 100.

The *South American Climate* (SAC) geodata stream[6] collects monthly-mean *air temperature* measurements (in Celsius degrees) recorded between 1960 and 1990 and interpolated over a 0.5° by 0.5° of latitude/longitude grid in South America. The grid nodes are centered on 0.25° for a total of 6477 sensors. The number of nearby stations that influence a grid-node estimate is 20 on average, which results in more realistic air-temperature fields. A sensor is considered spatially close to the sensors which are located in the cells around the grid. Regular and close-to periodic air temperature values range between −7.6 and 32.9.

The *Global Historical Climatology Network* geodata stream[7] (GHCN) collects monthly mean *air temperature* measurements (in Celsius degrees) for 7280 of land stations worldwide. The period of record varies from station to station, with several thousand extending back from 1890 up to 1999. The stations are unevenly installed around the world and the network configuration changes in time since new stations are installed at some time, while old stations are disused. A total of 1340 snapshots are collected. Both streams (in particular precipitation) include several missing values. A station is considered spatially close to the stations that are located in the range of two degrees longitude/latitude. By using a box plot, we find periodic and regular temperature values that presumably range between −20.75 and 49.25.

### 2.4.1.2  Evaluation Measures

Let $D$ be the geodata stream and $P$ be a summarization of $D$. The accuracy of $P$ in summarizing $D$ is evaluated by means of the root mean square error (*rmse*). Formally,

$$rmse(D, \hat{D}) = \sqrt{\frac{\sum_i (v_i - \hat{v}_i)^2}{|D|}}, \tag{2.23}$$

---

[5] http://db.csail.mit.edu/labdata/labdata.html

[6] http://climate.geog.udel.edu/~climate/html_pages/archive.html

[7] ftp://ftp.ncdc.noaa.gov/pub/data/ghcn/v2/

where $\hat{D}$ is the stream reconstructed from $P$. This error measures the deviation between original data before clustering them in trends and their prediction by means of the summarizing trend cluster stored in the database. Therefore, the lower the error, the more accurate the summarization. If the trend representation has been compressed before storage in the database, we use inverse transform to reconstruct the trend polyline to be used for predicting data.

The compression size (size$^{\%}$) is a value in percentage, which represents the ratio of the size of summary $P$ to the size of the original stream $D$, that is,

$$size^{\%}(D \rightarrow P) = \frac{size(P)}{size(D)} \times 100\%, \qquad (2.24)$$

with size$(\cdot)$ computed by taking into account that MySQL uses 2 bytes to store a SMALLINT (ranging between $-32767$ and $32767$) and 4 bytes to store a single precision FLOAT. The lower the size$^{\%}$, the more compact the $P$.

The average computation time per window is the time (in milliseconds) spent on average summarizing each window of $D$ and storing the summary in the database. SUMATRA can be considered a (near) real-time system if the time spent processing a window is less on average than the time spent buffering a new window. This aspect is remarkable in the evaluation of the IBL streams, where transmissions are very frequent (every 31 s).

## 2.4.2 Trend Cluster Analysis

We begin the evaluation study by investigating the summarization power of trend cluster discovery, without running any signal compression technique to compress trend polylines. We intend to study the influence of both window size $w$ and domain similarity threshold $\delta$ on the summarization power. Both $w$ and $\delta$ vary as reported in Table 2.1. For each geodata stream, $\delta$ ranges between 5, 10, and 20 % of the expected domain range of the measured attribute. Experiments with $w = 1$ are run to evaluate the quality of the summary if traditional spatial clusters (as reported in [17]) are used instead of trend clusters. The accuracy $(rmse)$, the average computation time per

**Table 2.1** SUMATRA parameter setting

| Stream | Dimension | $\delta$ | w |
|---|---|---|---|
| IBL | Temperature | 1.25, **2.5**, 5.0 | 1, 256, **512**,1024 |
| IBL | Humidity | 5, **10**, 20 | 1, 256, **512**, 1024 |
| SAC | Temperature | 2, **4**, 8 | 1, 6, **12**, 24 |
| GHCN | Temperature | 3.5, **7**, 14 | 1, 6, **12**, 24 |

Bold valued settings are those used to evaluate the compression polyline techniques

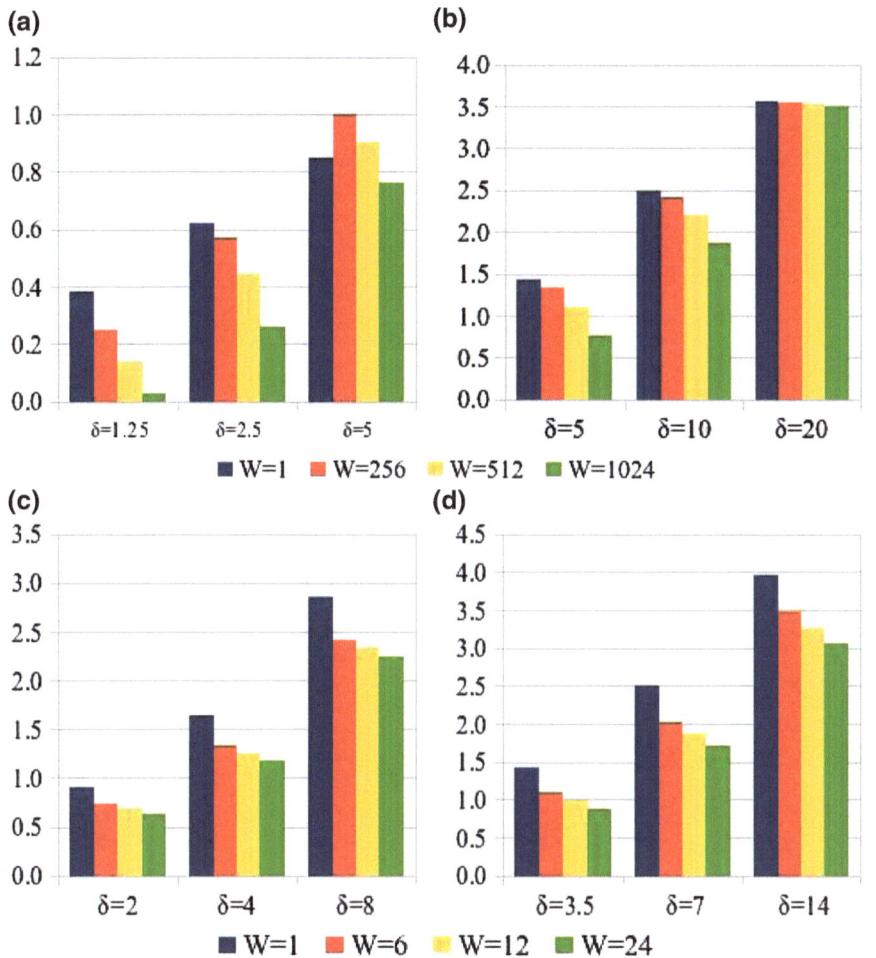

**Fig. 2.7** Trend cluster discovery: accuracy (rmse) is plotted (Y axis), by varying the window size (X axis). SUMATRA is run by varying the similarity domain threshold $\delta$. **a** IBL (Temperature). **b** IBL (Humidity). **c** SAC. **d** GHCN

window, and the compression size are plotted in Figs. 2.7, 2.8, 2.9 for the streams in this study. The analysis of these results leads to several considerations.

First, the root mean square error ($rmse$) is always significantly below $\delta$.

Second, trend clusters, discovered window-by-window ($w > 1$), generally summarize a stream better than spatial clusters, discovered snapshot-by-snapshot ($w = 1$). In particular, the accuracy obtained with the trend cluster summarization is greater than the accuracy obtained with the spatial cluster summarization. The general behavior which we observe is that by enlarging $w$ the accuracy of the summary increases.

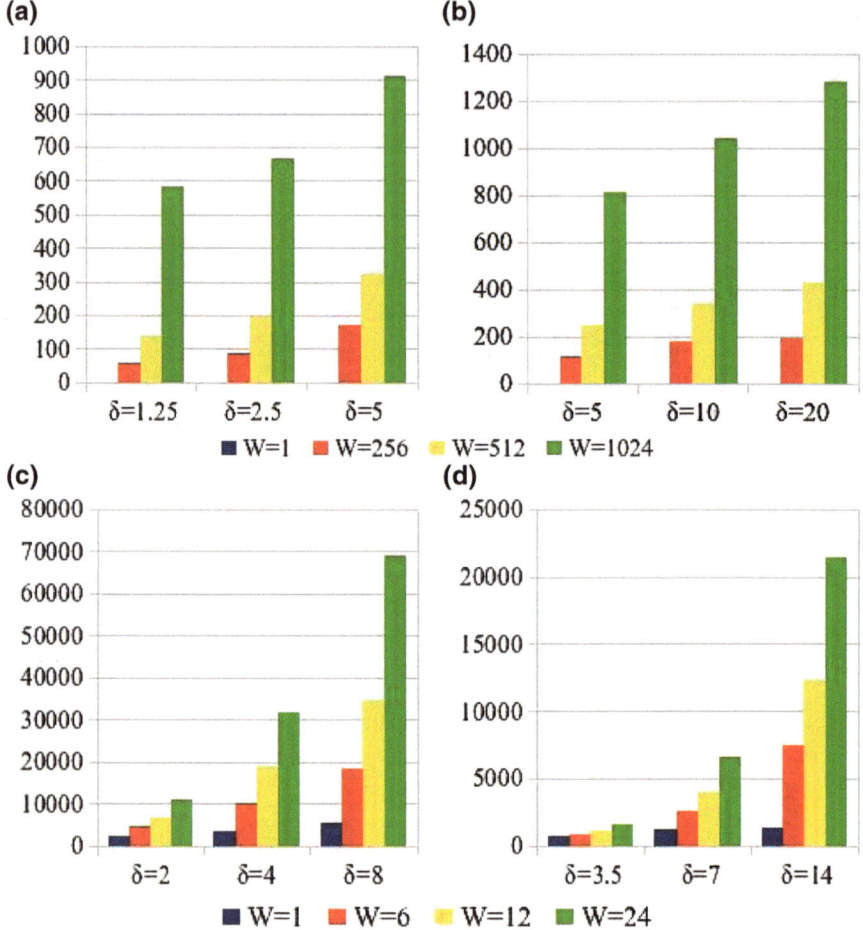

**Fig. 2.8** Trend cluster discovery: computation time spent per window (in milliseconds) is plotted (Y axis), by varying the window size (X axis). SUMATRA is run by varying the similarity domain threshold $\delta$. **a** IBL (Temperature). **b** IBL (Humidity). **c** SAC. **d** GHCN

This is due to the fact that the number of computed trend clusters tends to increase with $w$. The compression size of the trend clusters is always lower than 100 % and, in most cases, it is lower than the compression size of the spatial clusters. This does not happen only in a few cases, where $w$ is over-enlarged with respect to the number of nodes in the network (e.g., the Intel Berkeley Lab network). Further considerations are prompted by the analysis of the average computation time per window. As the computation time depends on the window size, computing trend clusters in a window is more time-consuming than computing spatial clusters in a snapshot. On the other hand, the total time spent processing the entire stream and computing spatial clusters at each snapshot is more than the time spent computing trend clus-

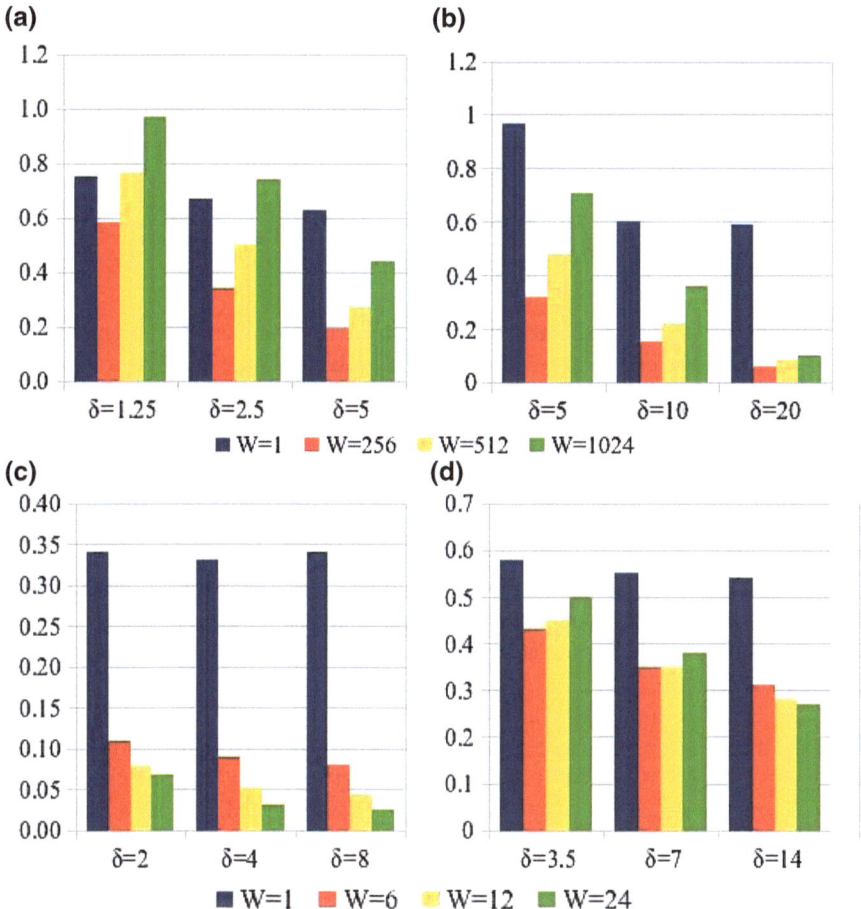

**Fig. 2.9** Trend cluster discovery: compression size is plotted (Y axis), by varying the window size (X axis). SUMATRA is run by varying the similarity domain threshold $\delta$. **a** IBL (Temperature). **b** IBL (Humidity). **c** SAC. **d** GHCN

ters (independently of the window size). In general, the analysis performed confirms our hypothesis that the use of trends in clustering sensors is an efficacious means for summarization purposes in terms of accuracy, size, and time complexity. However, at the same time it reveals that choosing the window size is a tricky step in SUMATRA. In any case, we point out here that the domain knowledge generally available on the nature of the monitored phenomenon (e.g., monthly temperature reasonably exhibits a year-long periodicity) plays a prominent role in this choice. In the absence of a domain-dependent guideline, a solution for automatically choosing the window size is monitoring the behavior of the summarizer for distinct window sizes along a time interval and then using the result of this comparative analysis to decide the window size to process the remaining stream.

**Table 2.2** SUMATRA versus ZIP technique: size of (compressed) data after storage in a standard text file

| Stream | Stream file (Mb) | Zip (Mb) | Trend cluster (Mb) | Trend cluster & Zip |
|---|---|---|---|---|
| IBL Temperature | 29.7 | 6.89 | 21.3 | 1.79 Mb |
| IBL Humidity | 29.9 | 6.45 | 9.39 | 744 Kb |
| SAC | 30.9 | 6.57 | 1.29 | 176 Kb |
| GHCN | 126 | 19.4 | 11.5 | 1.4 Mb |

Third, by increasing $\delta$, the accuracy of summarization decreases slightly but, as expected, the compression size is lower. Also in this case the average computation time per window increases with $\delta$, since the size of the neighborhoods explored to construct the clusters reasonably increases with $\delta$.

Fourth, experiments with IBL streams and GHCN stream confirm the capability of SUMATRA to deal with noise, outliers, and networks exhibiting a number of active sensors which are variable in time.

A final consideration concerns the analysis of the computation time. The computation time is always less than 1.25 s per window in the IBL streams, less than 70 s per window in the SAC stream and less than 25 s per window in the GHCN stream. These low values motivate our categorizing of SUMATRA as a system that processes data in real-time.

On the completion of this analysis, we compare SUMATRA with the straightforward standard zip file technique, which is commonly used to compress data files. To this end, we note that, even if a zip technique is indisputably very efficient and the original data can be exactly reconstructed from the zip bundle, it is purely frequency based and does not consider the spatiotemporal semantics of sensor data. On the contrary, the trend cluster summarization operates at a semantic level. Regarding this, our point of view is that a standard compression technique can be orthogonal to the trend cluster summarization, so that both techniques can be applied in sequence. To check the feasibility of this opinion, trend clusters, discovered with the parameter setting ($w$ and $\delta$) reported in bold in Table 2.1, have been stored in a standard text file rather than in a database. The size of this bundle is compared to the size of the zip bundle which zips the entire data stream, as well as the zip bundle which zips the trend clusters of a data stream. The analysis of the results, collected in Table 2.2, reveals that the data streams which are unaccounted for, due to missing data (SAC), or which have few missing data (GHCN), per window indisputably manifest a higher reduction capability of a spatiotemporal aware summarization technique. On the other hand, the computation of an interpolate, for which data are expected in a window, but which is missing in the stream (see Sect. **??** for details), allows the trend cluster summary to make an estimate of these missing data which is lacking in the zip bundle. The side effect observed from adding interpolated missing data to the summarization process is that the size of the summary is boosted. This explains why the trend cluster summary is larger than the zip bundle in streams where the percentage of missing data is very high (IBL). In any case, if the trend cluster summarization and the zip

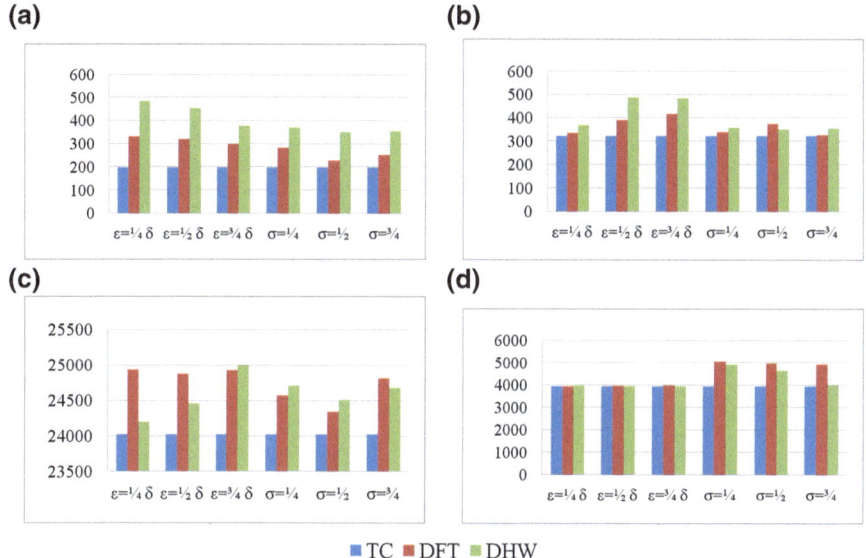

**Fig. 2.10**  Trend clusters (TC) versus Trend clusters+polyline compression (DFT, DHW): accuracy (rmse). **a** IBL (Temperature) $\delta = 2.5\ w = 512$. **b** IBL (Humidity) $\delta = 10\ w = 512$. **c** SAC $\delta = 4\ w = 12$. **d** GHCN $\delta = 7\ w = 12$

compression are applied in sequence, the size of the data storage is always reduced drastically.

### 2.4.3  Trend Compression Analysis

We consider the parameter setting ($w$ and $\delta$) reported in bold in Table 2.1 and evaluate the trend cluster summarization if the trend polyline representation is derived from DFT or DHW. Experiments are performed by computing the number ($k$) of coefficients, by fixing either the error threshold $\varepsilon$ or the compression degree threshold $\sigma$. For each stream, $\varepsilon$ ranges between $\frac{1}{4}\delta$, $\frac{1}{2}\delta$ and $\frac{3}{4}\delta$, while $\sigma$ ranges between $\frac{1}{4}$, $\frac{1}{2}$ and $\frac{3}{4}$. The average accuracy, computation time per window and compression size are plotted in Figs. 2.10, 2.11, 2.12 for each stream. Some considerations are reported in the following.

The integration of a trend polyline compression technique in SUMATRA has the expected effect of reducing even further the size of the summary, which was computed by the trend cluster discoverer. This result is achieved at the expense of accuracy and computation time. Additionally, the experimental results confirm that both tuning mechanisms are coherently defined. In fact, as expected, the use of the error-based tuning of $k$ leads to error ($rmse$) values which remain significantly

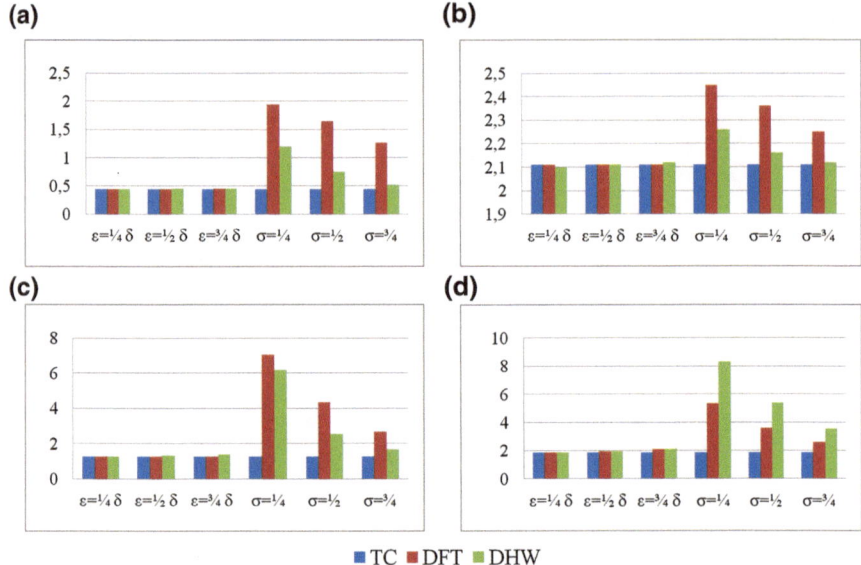

**Fig. 2.11** Trend clusters (TC) versus Trend clusters + polyline compression (DFT, DHW): computation time per window (in milliseconds). **a** IBL (Temperature) $\delta = 2.5\, w = 512$. **b** IBL (Humidity) $\delta = 10\, w = 512$. **c** SAC $\delta = 4\, w = 12$. **d** GHCN $\delta = 7\, w = 12$

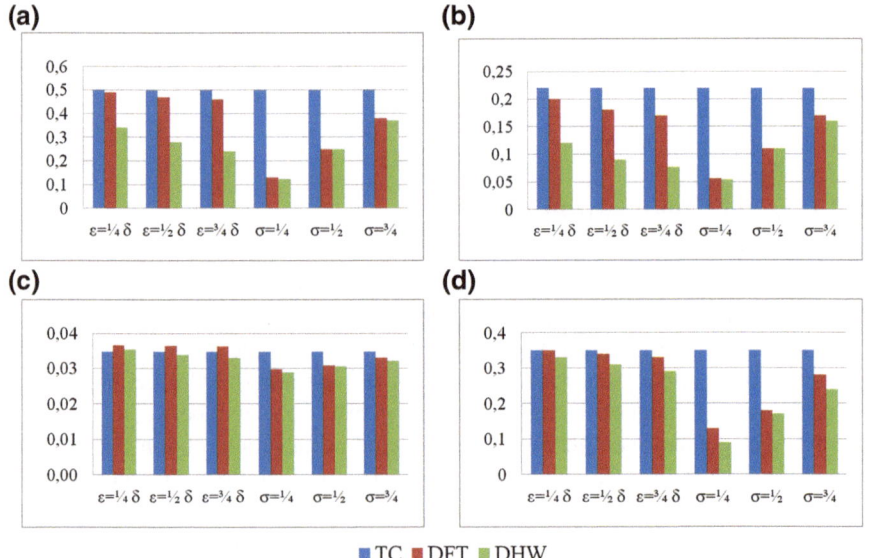

**Fig. 2.12** Trend clusters (TC) versus Trend clusters + polyline compression (DFT, DHW): compression size. **a** IBL (Temperature) $\delta = 2.5\, w = 512$. **b** IBL (Humidity) $\delta = 10\, w = 512$. **c** SAC $\delta = 4\, w = 12$. **d** GHCN $\delta = 7\, w = 12$

lower than $\delta + \varepsilon$. On the other hand, the size-based tuning of $k$ allows us to fix an upper bound for the size. In any case, by increasing $\varepsilon$ in the error-based tuning of $k$, the compared techniques show a reduction of the compression size but at a different speed (summaries become smaller). On the contrary, by increasing $\sigma$ in the size-based tuning of $k$, the techniques compared show a reduction in the errors (summaries become more accurate). In general, the lower the admitted size for the summary, the higher the error.

By comparing the signal compression technique, we observe that DHW always computes the smallest summary and it also achieves the highest accuracy in the majority of setups. This confirms the conclusions of the empirical study reported in [26].

Further notes are reported in this paragraph to explain the fact that the compression size does not benefit particularly, in percentage, from the trend polyline compression in both the South American Climate and the GHCN streams. In these streams, the network includes a large number of sensors, but the plausible window size and then the polyline size is relatively small ($w = 12$). Since most of the bytes are used to store clusters, the compression in the trend polyline representation does not produce the same significant reduction in the total size of the summary that we observe in the IBL streams.

The analysis of the average computation time per window shows how much the integration of a trend polyline compression technique in trend cluster discovery slows down the summarization process. In any case, the average computation time always remains competitively low; it is less than 0.5 s per window in the IBL streams, less than 25 s per window in the SAC stream and less than 6 s per window in the GHCN stream. This low computation time makes viable the performance of each trend polyline compression technique within the trend clustering discovery process, without undermining the real-time behavior of SUMATRA.

## 2.5 Trend Cluster-Based Data Cube

The trend cluster discovery theory computes aggregates on geodata streams by considering time and space as aggregation dimensions. This paves the way to extend the cube technology [27] to geodata, in order to naturally organize the stream storage around time and space dimensions [28].

### 2.5.1 Geodata Cube

We describe now how trend clusters can represent an appropriate aggregation model for a geodata stream cube architecture.

**Definition 2.9** (**Geodata Cube**) A geodata cube $\mathscr{Q}$ is the triple:

$$(Z, D, F), \tag{2.25}$$

where

1. $Z$ is the cube measure (attribute measured through a sensor network $K$);
2. $D$ is the non-empty set of dimensional attributes, which includes the time line $T$ and a time-based space function $space(t)$ with $t \in T$;
3. $F$ is the unbounded fact table populated with the snapshots of the geodata stream $z(T, K)$.

Every dimensional attribute of the cube is associated to a hierarchy of levels such that each level is a set of dimension values and there exists a partial order based on a containment relation ($\sqsupseteq$) according to which, for two levels in a dimension, a value at the higher level contains a set of values at the lower level. The structure of both hierarchies is defined in the following for the count-based model of a stream.

**Definition 2.10** (**Time Hierarchy** $\mathscr{H}(T)$) Let:

1. $w$ be the size of a count-based model;
2. $\Omega$ be a window multiplier.

The hierarchy $\mathscr{H}(T)$ is defined, depending on $w$ and $\Omega$, by a containment relation formulated as follows:

$$\underbrace{T}_{time\ line} \sqsupseteq \dots \sqsupseteq \underbrace{t_{((i-1)\ \mathrm{mod}\ \Omega w)\Omega w+1} \to t_{((i-1)\ \mathrm{mod}\ \Omega w+1)\Omega w}}_{higher\ level\ window} \sqsupseteq \underbrace{t_{(i-1)w+1} \to t_{iw}}_{window} \sqsupseteq \underbrace{t}_{time\ point}, \tag{2.26}$$

where $t$ is a specific time point of $T$, which timestamps one of the snapshots of the stream. $t_{(i-1)w+1} \to t_{iw}$ is the time horizon of a window in the count-based model having size $w$. $t_{((i-1)\ \mathrm{mod}\ \Omega w)\Omega w+1} \to t_{((i-1)\ \mathrm{mod}\ \Omega w+1)\Omega w}$ is the time horizon of the higher level window in the count-based stream model with size $w\Omega$.

**Definition 2.11** (**Space Hierarchy** $\mathscr{H}(space(T))$) Let:

1. $t$ be a time point of $T$ and $t_i \to t_e$ be the time horizon of a window of a count-based model of $z(K, T)$, such that $t \geq t_i$ and $t \leq t_e$;
2. $\delta$ be the trend similarity threshold such that $\mathscr{P}(\mathscr{C})$ is the set of spatial clusters of trend clusters discovered in $t_i \overset{z(T,K)}{\to} t_e$ with threshold $\delta$;
3. $\Delta$ be a cluster multiplier.

The hierarchy $\mathscr{H}(space(t))$ is defined for the time point $t$, depending on $\delta$ and $\Delta$, by the containment relation formulated as follows:

$$\underbrace{space[W]}_{space} \sqsupseteq \dots \sqsupseteq \underbrace{\mathscr{C}^{\Delta}}_{higher\ level\ cluster} \sqsupseteq \underbrace{\mathscr{C}}_{cluster} \sqsupseteq \underbrace{[x, y]}_{spatial\ point}, \tag{2.27}$$

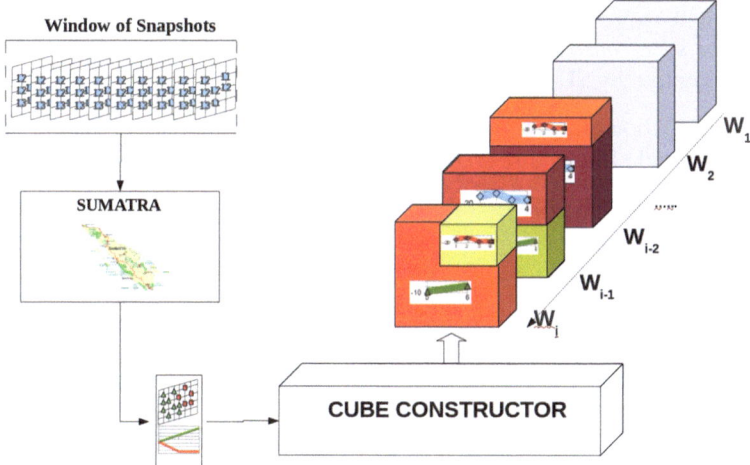

**Fig. 2.13**  Geo-Trend Stream Cube

where $[x, y]$ is the spatial location of a sensor of $K$. $\mathscr{C}$ is a cluster of $\mathscr{P}(\mathscr{C})$, $\mathscr{C}^{\Delta}$ is the higher level cluster, which groups spatially close clusters (i.e., clusters which group spatially close sensors), which have trend polylines differing one from each other at worst $\Delta$.

## 2.5.2 Stream Cube Creation

GeoTube (GEO-Trend stream cUBEr) provides facilities for computing, storing and exploring a geodata stream cube. The cube computation can be triggered by the following statement:

```
CREATE GEOTRENDCUBE 𝒬 WITH MEASURE Z FROM STREAM z(T, K)
GROUPING BY SPACE, TIME
HAVING SIMILARITY δ AND SIZE w
```

The cube $\mathscr{Q}$ is computed from the geodata stream $z(T, K)$ and stored permanently in a database. The time horizon of the windows of the $w$-sized count-based stream model determines the aggregation level for the time, while the spatial clusters, which are computed by the $\delta$-aware trend cluster discovery process, determine the aggregation level for the space.

The architecture of the GeoTube (see Fig. 2.13) comprises three components, that is,

1. a *snapshot buffer*, which consumes data snapshots as they arrive and pours them window-by-window into SUMATRA;

2. the system *SUMATRA*, which performs the trend cluster discovery process; and
3. the *cube slice constructor*, which builds a new slice of the cube by using trend clusters output by SUMATRA.

The construction of a cube slice proceeds as follows. Let $P(t_{(i-1)w+1} \rightarrow t_{iw})$ be the set of $k$ trend clusters discovered with trend similarity threshold $\delta$ from the $i$th $w$-sized data window $t_{(i-1)w+1} \overset{z(T,K)}{\rightarrow} t_{iw}$. The cube slice $\mathscr{Q}[t_{(i-1)w+1} \rightarrow t_{iw}]$,

$$\mathscr{Q}[t_{(i-1)w+1} \rightarrow t_{iw}] = \begin{array}{|c|c|c|} \hline \text{time} & \text{space} & \text{measure} \\ \hline t_{(i-1)w+1} \rightarrow t_{iw} & \mathscr{C}_1 & \mathscr{Z}_1 \\ \hline t_{(i-1)w+1} \rightarrow t_{iw} & \ldots & \ldots \\ \hline t_{(i-1)w+1} \rightarrow t_{iw} & \mathscr{C}_k & \mathscr{Z}_k \\ \hline \end{array} \tag{2.28}$$

can be defined by considering each trend cluster $[t_{(i-1)w+1} \rightarrow t_{iw}, \mathscr{C}, \mathscr{Z}] \in P(t_{(i-1)w+1} \rightarrow t_{iw})$.

The existing time hierarchy $\mathscr{H}(T)$ is expanded. Time points, which timestamp snapshots acquired between $t_{(i-1)w+1}$ and $t_{iw}$, are added to the bottom level of $\mathscr{H}(T)$. The time horizon $t_{(i-1)w+1} \rightarrow t_{iw}$ is added, as a grouping value of time, at the window level of $\mathscr{H}(T)$.

A new space hierarchy $\mathscr{H}(space(t))$ is created. It is time-based defined with time $t$ ranging between $t_{(i-1)w}$ and $t_{iw}$. Spatial points, which georeference sensors of $K$, are added to the bottom level of $\mathscr{H}(space(t))$. Each cluster $\mathscr{C}$ of the trend cluster set $P(t_{(i-1)w+1} \rightarrow t_{iw})$ is added, as the grouping value of the space, at the cluster level of $\mathscr{H}(space(t))$.

For each cluster $\mathscr{C}$ of the trend cluster set $P(t_{(i-1)w+1} \rightarrow t_{iw})$, the trend polyline prototype $\mathscr{Z}$ associated to $\mathscr{C}$ is stored in the cube cell $\mathscr{Q}[t_{(i-1)w+1} \rightarrow t_{iw}][\mathscr{C}]$.

It is noteworthy that, due to the stream compression naturally operated by trend clusters, the memory size for storing $\mathscr{Q}$ grows indefinitely, but less than the memory size for storing of the stream $z(T, K)$.

### 2.5.3 Roll-up

The roll-up request can be formulated as follows:

```
ROLL-UP on MEASURE Z OF GEOTRENDCUBE 𝒞
WITH SPACE Δ and TIME Ω
```

and triggers the roll-up from cube $\mathscr{Q}$ to cube $\mathscr{Q}'$ such that:

1. cube $\mathscr{Q}'$ has the same base stream as $\mathscr{Q}$;
2. the time hierarchy $\mathscr{H}'(T)$ groups the time points of $T$ on the windows of the $(w\Omega)$-sized count-based model of the base stream;

3. for each window of the $(w\Omega)$-sized count-based stream model, the associated space hierarchy $\mathcal{H}'(space(t))$ is populated according to the trend clusters, which group the data of the window with trend similarity threshold $\delta + \Delta$.

The procedure to roll-up $\mathscr{C}$ into $\mathscr{C}'$ is iterative. It processes cube slices of $\mathscr{Q}$ by going along the time line $T$ with step $\Omega$. At each iteration $i$, it builds the cube slice $\mathscr{Q}'[t_{(i-1)\Omega w+1} \to t_{i\Omega w}]$ by the $i$th series of $\Omega$ consecutive cube slices of $\mathscr{Q}$, that is,

$$\mathscr{Q}[t_{(i-1)\Omega w+1} \to t_{(i-1)\Omega w+w}]$$
$$\mathscr{Q}[t_{(i-1)\Omega w+w+1} \to t_{(i-1)\Omega w+2w}]$$
$$\mathscr{Q}[t_{(i-1)\Omega w+2w+1} \to t_{(i-1)\Omega w+3w}]$$
$$\cdots$$
$$\mathscr{Q}[t_{(i-1)\Omega w+(\Omega-1)w+1} \to t_{i\Omega w}]$$

The construction of $\mathscr{Q}'[t_{(i-1)\Omega w+1} \to t_{(i)\Omega w}]$ proceeds as follows. For each $\mathscr{Q}[t_{(i-1)\Omega w+(j-1)w+1} \to t_{(i-1)\Omega w+jw}]$ (with $j$ ranging between 1 and $\Omega$), the associated trend cluster set $P(t_{(i-1)\Omega w+(j-1)w+1} \to t_{(i-1)\Omega w+jw})$ is retrieved from $\mathscr{Q}$. This is done by retrieving:

1. each value $\mathscr{C}$, which appears at the cluster level of the space hierarchy $H(space(t))$ with $t$ between $t_{(i-1)\Omega w+(j-1)w+1}$ and $t_{(i-1)\Omega w+jw}$;
2. for each cluster value $\mathscr{C}$, the trend polyline prototype $\mathscr{Z}$, which is stored in the time-space cube cell $\mathscr{Q}[t_{(i-1)\Omega w+(j-1)w+1} \to t_{(i-1)\Omega w+jw}][\mathscr{C}]$.

This series of trend cluster sets:

$$P(t_{(i-1)\Omega w+1} \to t_{(i-1)\Omega w+w})$$
$$P(t_{(i-1)\Omega w+w+1} \to t_{(i-1)\Omega w+2w})$$
$$\cdots$$
$$P(t_{(i-1)\Omega w+(\Omega-1)w+1} \to t_{i\Omega w})$$

is rolled-up into the trend cluster set $P'(t_{(i-1)\Omega w+1} \to t_{i\Omega w})$, such that:

1. the time horizon $(t_{(i-1)\Omega w+1} \to t_{i\Omega w})$ is that of the $i$th window of the $\Omega w$-sized count-based stream model;
2. the spatial cluster set $\{\mathscr{C}'\}$ associated to $P'$ groups the sensors of $K$ around trend polyline prototypes $\mathscr{Z}'$ with length $\Omega w$ and trend similarity threshold $\delta + \Delta$.

Finally, the cube slice constructor inputs $P'(t_{(i-1)\Omega w+1} \to t_{i\Omega w})$ to populate the cube slice $\mathscr{Q}'(t_{(i-1)\Omega w+1} \to t_{i\Omega w})$ (details in Sect. 2.5.2).

The roll-up algorithm, which computes $P'$, is reported in Algorithm 2.3. It is two stepped.

SPACE ROLL-UP [lines 2–4, Algorithm 2.3]

---

**Algorithm 2.3** ROLL-UP($\{P(t_{(i-1)\Omega w+(j-1)w+1} \rightarrow t_{(i-1)\Omega w+jw})\}_{j=1,\ldots,\Omega}, \Omega, \Delta)$
$\mapsto P'(t_{(i-1)\Omega w+1} \rightarrow t_{i\Omega w})$

---

**Require:** $P(t_{(i-1)\Omega w+(j-1)w+1} \rightarrow t_{(i-1)\Omega w+jw})$: the trend cluster set with time horizon
$t_{(i-1)\Omega w+(j-1)w+1} \rightarrow t_{(i-1)\Omega w+jw}$
**Require:** $\Omega$: the grouping factor for the ROLL-UP in time
**Require:** $\Delta$: the grouping factor for the ROLL-UP in space
**Ensure:** $P'(t_{(i-1)\Omega w+1} \rightarrow t_{i\Omega w})$: the trend cluster set with time horizon $t_{(i-1)\Omega w+1} \rightarrow t_{i\Omega w}$
1: $P'(t_{(i-1)\Omega w+1} \rightarrow t_{i\Omega w}) \leftarrow \oslash$
    Roll-Up in SPACE
2: **for all** $j = 1$ to $\Omega$ **do**
3:    $\widetilde{P}(t_{(i-1)\Omega w+(j-1)w+1} \rightarrow t_{(i-1)\Omega w+jw}) \leftarrow$ SUMATRA($P(t_{(i-1)\Omega w+(j-1)w+1} \rightarrow$
     $t_{(i-1)\Omega w+jw}), \Delta)$
4: **end for**
    Roll-up in TIME
5: **for all** ($s \in K$ and $s$ is unclustered) **do**
6:    $\mathscr{C}' \leftarrow \{s\}$;
7:    $\mathscr{C}' \leftarrow$ expandROLLUPCluster($s, \mathscr{C}', \{\widetilde{\mathscr{C}}_j^{[s]}\}_{j=1,\ldots,\Omega}$)
8:    $\mathscr{Z}' \leftarrow \widetilde{\mathscr{Z}}_1^{[s]} \bullet \ldots \bullet \widetilde{\mathscr{Z}}_\Omega^{[s]}$;
9:    add($P'(t_{(i-1)\Omega w+1} \rightarrow t_{i\Omega w}), \{(t_{(i-1)\Omega w+1} \rightarrow t_{i\Omega w}, \mathscr{C}', \mathscr{Z}')\})$
10: **end for**

---

Each input trend cluster set $P(t_{(i-1)\Omega w+(j-1)w+1} \rightarrow t_{(i-1)\Omega w+jw})$ is processed: trend polyline prototypes, differing at worst $\Delta$ from each other, are clustered into a single trend cluster.

We use SUMATRA to perform this clustering process (line 3, Algorithm 2.3). The discovery process is run with trend similarity threshold $\Delta$ by considering each cluster of the cluster set of $P(t_{(i-1)\Omega w+(j-1)w+1} \rightarrow t_{(i-1)\Omega w+jw})$ as a single sensor and the associated trend polyline prototype as the series of measurements transmitted by the cluster source.

An edge relation exists between two cluster sources if there exists at least one edge relation between two sensors belonging to these clusters.

The output is denoted by $\widehat{P}(t_{(i-1)\Omega w+(j-1)w+1} \rightarrow t_{(i-1)\Omega w+jw})$ (see Fig. 2.14).

TIME ROLL-UP [lines 5–10, Algorithm 2.3]

Sensors, which are repeatedly classified together in a cluster along the time horizon $t_{(i-1)\Omega w+(j-1)w+1} \rightarrow t_{(i-1)\Omega w+jw}$, are searched for. They measure data, which evolve with a similar trend prototype along the considered time horizon. Hence, they can be clustered in a single trend cluster with time horizon $t_{(i-1)\Omega w+1} \rightarrow t_{i\Omega w}$ (see Fig. 2.15).

The construction of a trend cluster $(t_{(i-1)\Omega w+1} \rightarrow t_{i\Omega w}, \mathscr{C}', \mathscr{Z}')$ starts by choosing the sensor $s$ not yet clustered (line 5, Algorithm 2.3).

Let $(t_{(i-1)\Omega w+(j-1)w+1} \rightarrow t_{(i-1)\Omega w+jw}, \widetilde{\mathscr{C}}_j^{[s]}, \widetilde{\mathscr{Z}}_j^{[s]})$ be a trend cluster of the set $\widetilde{P}(t_{(i-1)\Omega w+(j-1)w+1} \rightarrow t_{(i-1)\Omega w+jw})$, such that $\mathscr{C}^{[s]}$ contains $s$. The new cluster $\mathscr{C}'$, which initially contains $s$ (line 6, Algorithm 2.3), is expanded (line 7, Algorithm 2.3).

**Algorithm 2.4** expandROLLUPCluster$(s, \mathscr{C}', \{\widetilde{\mathscr{C}}_j^{[s]}\}_{j=1,\ldots,\Omega}) \mapsto \mathscr{C}'$

---

1: **for all** $(s' \in K$ with $s'$ spatially close to $s$ and $s$ unclustered$)$ **do**
2:    **if** isCluster$(s', \{\widetilde{\mathscr{C}}_j^{[s]}\}_{j=1,\ldots,\Omega})$ **then**
3:       $\mathscr{C}' \leftarrow$ expandCluster$(s', \mathscr{C}' \cup \{s\}, \{\widetilde{\mathscr{C}}_j^{[s]}\}_{j=1,\ldots,\Omega})$;
4:    **end if**
5: **end for**

---

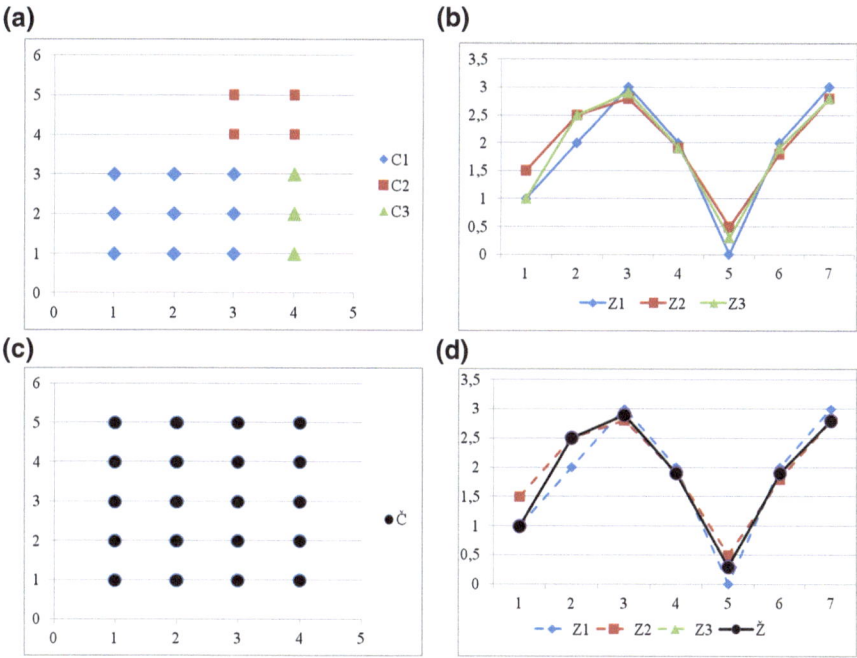

**Fig. 2.14** Space Roll-up: The input trend clusters (C1, Z1), (C2, Z2), and (C3, Z3) are grouped in the new trend cluster (C', Z') with $\Delta = 0.3$. **a** Clusters. **b** Trends. **c** Rolled-up Clusters. **d** Rolled-up Trends

The cluster expansion (lines 1–5, Algorithm 2.4) is performed by expanding $\mathscr{C}'$ with the unclustered sensors $s'$, which are in the neighborhood of the seed $s$ and are classified as $s$ in the window $(\widetilde{\mathscr{C}}_j^{[s]} = \widetilde{\mathscr{C}}_j^{[s']}$ for each $j$ ranging between 1 and $\Omega$).

The cluster expansion process is repeated by considering each sensor point, already grouped in $\mathscr{C}'$, as an expansion seed (line 3, Algorithm 2.4). If no sensor can be added to $\mathscr{C}'$, $\mathscr{Z}'$ is built by sequencing the trend prototypes $\widetilde{\mathscr{Z}}_j^{[s]}$ with $j = \ldots \Omega$ (line 8, Algorithm 2.3).

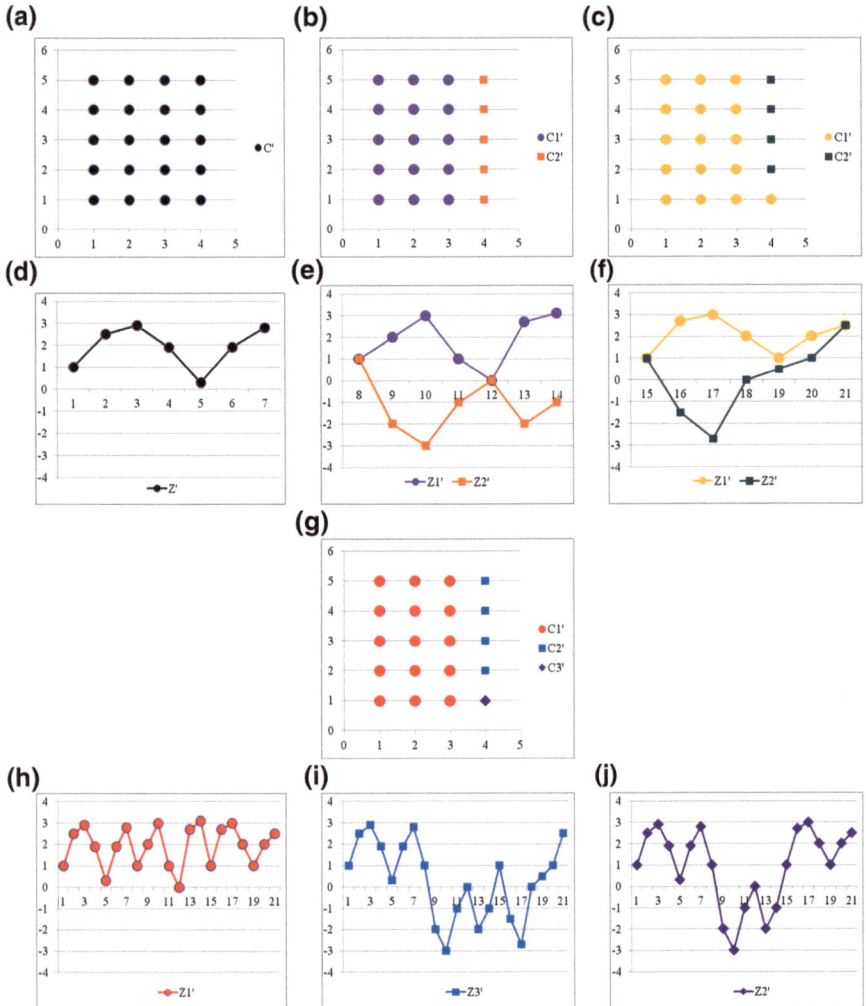

**Fig. 2.15** Time Roll-up: Trend clusters are created by grouping sensors which are repeatedly clustered together at $\Omega$ consecutive windows ( with $\Omega = 3$). Trends are obtained by sequencing the $\Omega$ trends in the associated windows. **a** Clusters $t_1 \rightarrow t_7$. **b** Clusters $t_8 \rightarrow t_{14}$. **c** Clusters $t_{15} \rightarrow t_{21}$. **d** Trends $t_1 \rightarrow t_7$. **e** Trends $t_8 \rightarrow t_{14}$. **f** Trends $t_{15} \rightarrow t_{21}$. **g** Rolled-up Clusters $t_1 \rightarrow t_{21}$. **h** Rolled-up Trends $t_1 \rightarrow t_{21}$. **i** Rolled-up Trends $t_1 \rightarrow t_{21}$. **j** Rolled-up Trends $t_1 \rightarrow t_{21}$

## 2.5.4 Drill-Down

The drill-down request is formulated as follows:

```
DRILL-DOWN on MEASURE Z OF GEOTRENDCUBE 𝒬
```

---

**Algorithm 2.5** DRILL-DOWN($P(t_{(i-1)w+1} \rightarrow t_{iw})) \mapsto \{P'(t_{(i-1)w+j} \rightarrow t_{(i-1)w+j})\}_{j=1,\ldots,w}$

---

1: **for all** ($j = 1$ **to** $w$) **do**
2:     $P'(t_{(i-1)w+j} \rightarrow t_{(i-1)w+j}) = \oslash$
3:     **for all** ($s \in K$) **do**
4:         $(t_{(i-1)w+1} \rightarrow t_{iw}, \mathscr{C}^{[s]}, \mathscr{X}^{[s]}) \leftarrow$ trendcluster($s, P(t_{(i-1)w+1} \rightarrow t_{iw})$)
5:         add($P'(t_{(i-1)w+j} \rightarrow t_{(i-1)w+j}), \{(t_{(i-1)w+j} \rightarrow t_{(i-1)w+j}, \{s\}, [(j, \mathscr{X}^{[s]}[j])]\}$)
6:     **end for**
7: **end for**

---

and triggers the drill-down process from $\mathscr{Q}$ to $\mathscr{Q}'$, such that:

1. cube $\mathscr{Q}'$ has the same base stream as $\mathscr{Q}$;
2. the time hierarchy $\mathscr{H}'(T)$ groups the time points of $T$ on windows of the 1-sized count-based stream model;
3. for each widow of the 1-sized count-based model, the associated space hierarchy $\mathscr{H}'(space(t))$ is populated by using the trend clusters, which cluster a single point.

The procedure to drill-down from $\mathscr{C}$ to $\mathscr{C}'$ is iteratively defined. At each iteration $i$, it processes the $i$th cube slice of $\mathscr{Q}$ by going along the time line $T$.

Let $P(t_{(i-1)w+1} \rightarrow t_{iw})$ be the trend cluster set associated to the slice $\mathscr{Q}[t_{(i-1)w+1} \rightarrow t_{iw}]$. $w$ trend cluster sets, $P'(t_{(i-1)w+j} \rightarrow t_{(i-1)w+j})$, with $j$ ranging between 1 and $w$, are computed and input to the cube slice constructor to populate $\mathscr{Q}'$ (see details in Sect. 2.5.2).

The computation of each $P'(t_{(i-1)w+j} \rightarrow t_{(i-1)w+j})$ proceeds as reported in Algorithm 2.5. For each sensor point $s \in K$ (line 3, Algorithm 2.5), the trend cluster $(t_{(i-1)w+1} \rightarrow t_{iw}, \mathscr{C}^{[s]}, \mathscr{X}^{[s]})$, which clusters $s$, is identified (i.e., $s \in \mathscr{C}^{[s]}$). For each $j$ ranging between 1 and $w$, the trend cluster $(t_{(i-1)w+j} \rightarrow t_{(i-1)w+j}, \{s\}, \mathscr{X}^{[s]}[j])$ is output.

## 2.5.5 A Case Study

We describe an application of GeoTube to maintain the electrical power (in kw/h) weekly transmitted from PhotoVoltaic (PV) plants. The stream is generated with PVGIS[8] (http://re.jrc.ec.europa.eu/pvgis) by distributing 52 PV plants over the South of Italy. Each plant is 0.5° of latitude/longitude distance apart from the others. The weekly estimates of electricity production are obtained by the default parameter setting in PVGIS and streamed for 52 weeks.

The GeoTube slice constructor is used to store the stream in cube $\mathscr{Q}$ computed with $w = 4$ and $\delta = 1.5$ Kw/h. The graph structure for the clustering phase is defined

---

[8] PVGIS is a map-based inventory of the PV plants electricity productions.

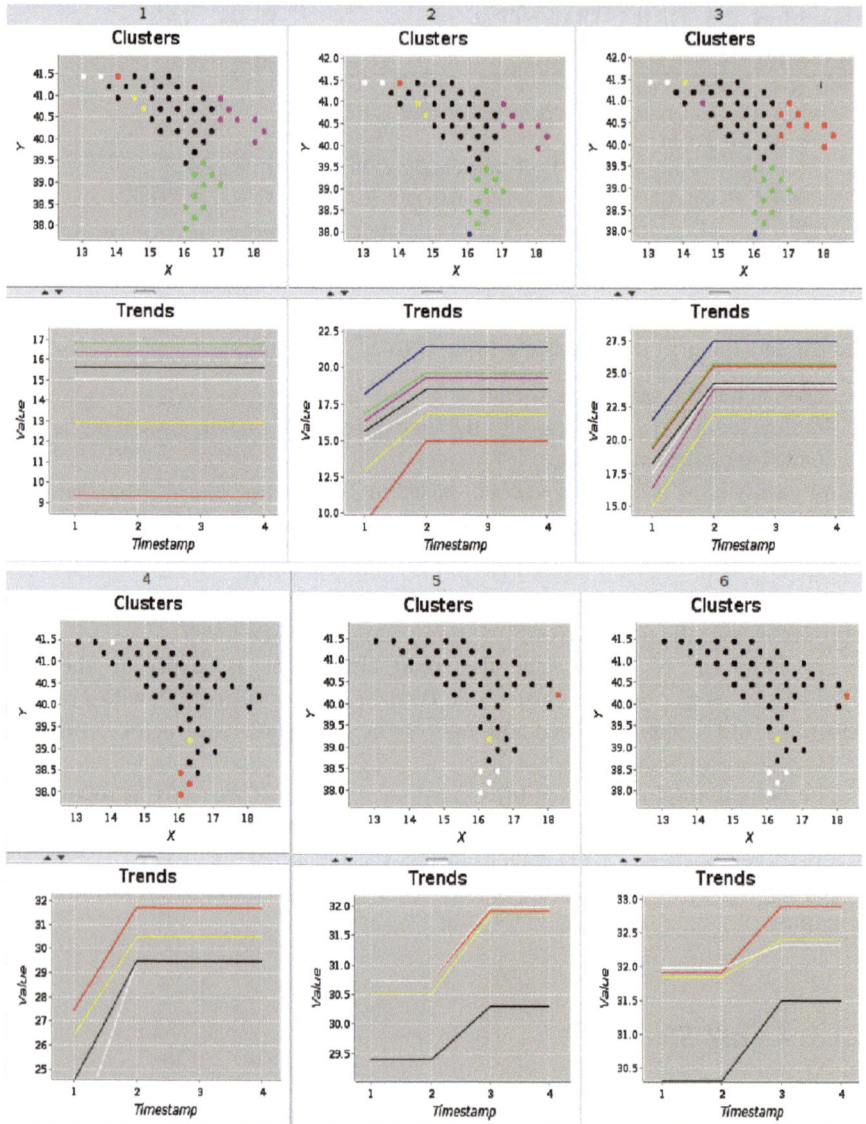

**Fig. 2.16** Geo-Trend Stream Cube construction with $w = 4 - \delta = 1.5$ Kw/h: windows 1–6

by the distance relation between the plants, i.e., 2 PV plants are edged if the distance between them is less than $0.5°$ of latitude/longitude.

The cube slices of $\mathcal{Q}$ are plotted in Figs. 2.16, 2.17.

We can navigate $\mathcal{Q}$ one level up by rolling-up in time with $\Omega = 3$ and in space with $\Delta = 1.5$. In this way, we are able to explore data streams (see Fig. 2.18) by

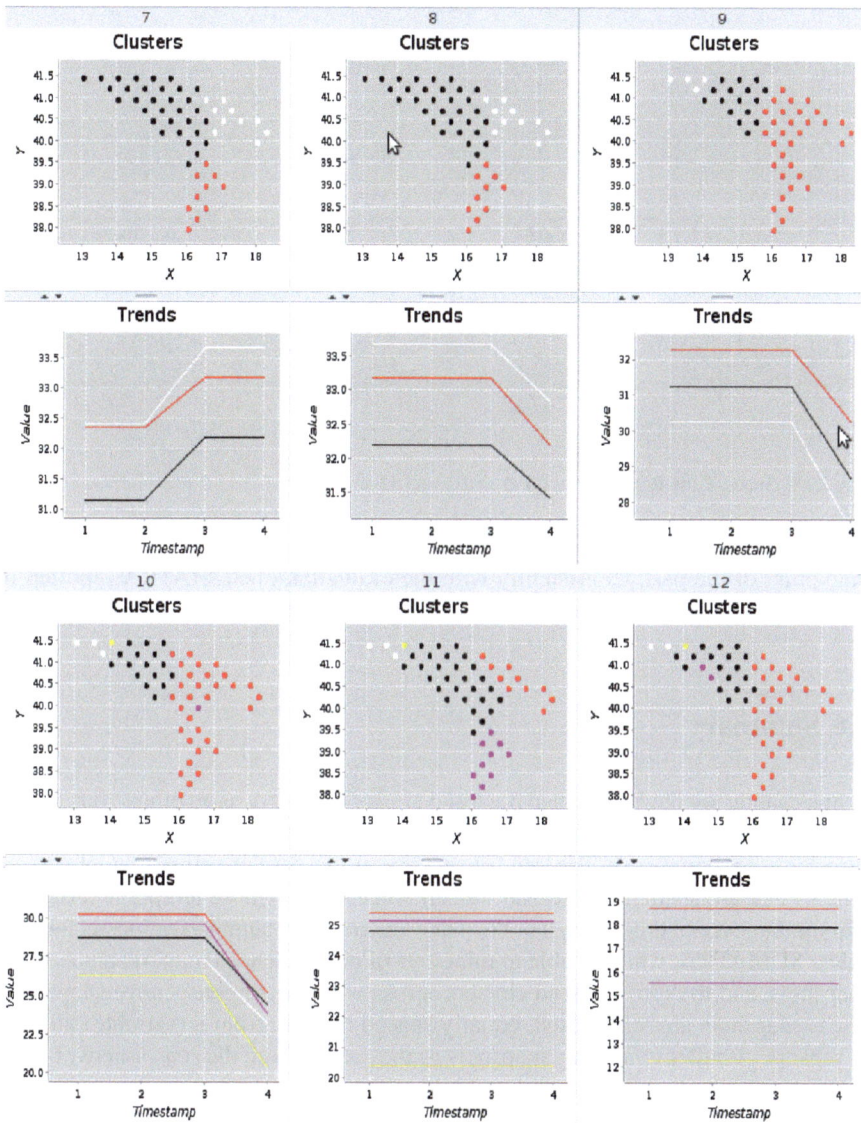

**Fig. 2.17** Geo-Trend Stream Cube construction with $w = 4 - \delta = 1.5$ Kw/h: windows 7–12

looking for longer trends (12 weeks), shared by larger clusters. It is noteworthy that if $\Omega = 1$ and $\Delta > 0$, then the roll-up is only in space. On the other hand, if $\Omega > 1$ and $\Delta = 0$, then the roll-up is only in time.

Finally, we can navigate $\mathcal{Q}$ one level down by drilling-down both in time and in space. In this way, we use DRILL-DOWN($\mathscr{C}$) to query stream data for any specific

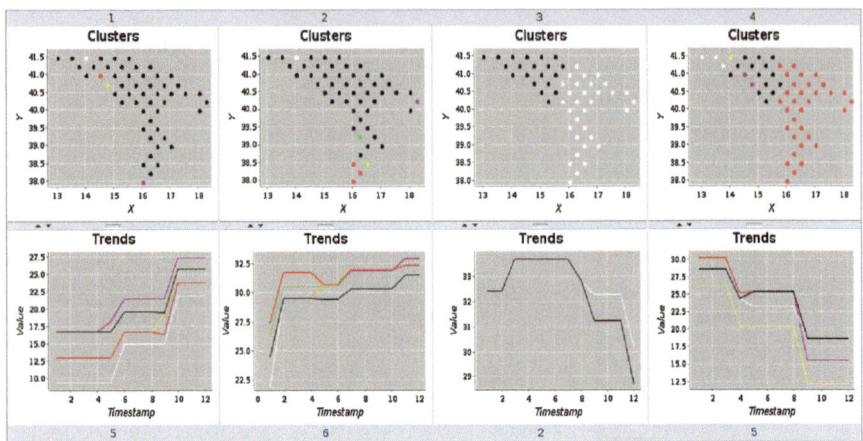

**Fig. 2.18** Space-Time ROLL-UP with $\Delta = 1.5$ and $\Omega = 3$

time point of the past. Consistently with the evaluation of SUMATRA reported in Sect. 2.4, the root mean square error is always under 0.5 ($<\delta$).

## 2.6 Summary

In this chapter we have described the trend cluster discovery as an efficacious means for addressing the task of summarization for a geodata stream produced from a (wireless) sensor network. The trend cluster is a special case of a cluster which groups sensors that are spatially close and transmits measures, whose temporal variations are similar over a time horizon. We have described a summarization technique, called SUMATRA, which is able to mine sets of georeferenced numeric data, called snapshots, and to discover trend clusters across windows of consecutive snapshots. These snapshots are transmitted, equally spaced in time, from a (variable) number of sensors. A buffer consumes snapshots as they arrive from the sensor network and pours them, window-by-window, into SUMATRA. Trend clusters are then computed as the summarization of each window. The window is discarded, while a compact representation of trend cluster polylines is computed and stored in a database along with the cluster. We observe that trend cluster discovery can be compared to the roll-up of windows for storage in a data warehouse. Drill-down is always possible to approximately reconstruct the stream from these summaries. For each past window, the values of the trend polylines of this window are forecast for each sensor grouped in the associated cluster. SUMATRA also allows us to obtain a compact representation of a trend polyline, by resorting to signal processing techniques. Both frill-down and roll-up operations are supported by the system GeoTube which used trend clusters to build a geo-trend stream cube.

# References

1. R. Motwani, P. Raghavan, *Randomized Algorithms* (Cambridge University Press, New York, 1995)
2. S. Acharya, P.B. Gibbons, V. Poosala, Congressional samples for approximate answering of group-by queries, in *Proceedings of the International Conference on Management of Data (SIGMOD 2000)* (ACM, 2000), pp. 487–498
3. Y. Zhu, D. Shasha, Statstream: statistical monitoring of thousands of data streams in real time, in *Proceedings of the 28th International Conference on Very Large Data Bases (VLDB 2002)* (VLDB Endowment, 2002), pp. 358–369
4. H.V. Jagadish, N. Koudas, S. Muthukrishnan, V. Poosala, K.C. Sevcik, T. Suel, Optimal histograms with quality guarantees, in *Proceedings of the 24th International Conference on Very Large Data Bases (VLDB 1998)* (Morgan Kaufmann, 1998), pp. 275–286
5. A.C. Gilbert, S. Guha, P. Indyk, Y. Kotidis, S. Muthukrishnan, M.J. Strauss, Fast, small-space algorithms for approximate histogram maintenance, in *Proceedings of the 24th Annual ACM Symposium on Theory of Computing (STOC 2002)* (ACM, 2002), pp. 389–398
6. M. Greenwald, S. Khanna, Space-efficient online computation of quantile summaries. ACM SIGMOD Rec. **30**(2), 58–66 (2001)
7. Y.E. Ioannidis, V. Poosala, Balancing histogram optimality and practicality for query result size estimation, in *Proceedings of the International Conference on Management of Data (SIGMOD 1995)* (ACM, 1995), pp. 233–244
8. N. Thaper, S. Guha, P. Indyk, N. Koudas, Dynamic multidimensional histograms, in *Proceedings of the International Conference on Management of Data (SIGMOD 2002)* (ACM, 2002), pp. 428–439
9. F. Furfaro, G.M. Mazzeo, D. Saccà, C. Sirangelo, Compressed hierarchical binary histograms for summarizing multi-dimensional data. Knowl. Inf. Syst. **15**(3), 335–380 (2008)
10. N. Alon, Y. Matias, M. Szegedy, The space complexity of approximating the frequency moments, in *Proceedings of the 28th Annual ACM Symposium on Theory of Computing (STOC 1996)*, (ACM, 1996), pp. 20–29
11. F. Rusu, A. Dobra, Sketching sampled data streams, in *Proceedings of the 25th International Conference on Data Engineering (ICDE 2009)* (IEEE Computer Society, 2009), pp. 381–392
12. J. Hershberger, N. Shrivastava, S. Suri, C.D. Toth, Adaptive spatial partitioning for multidimensional data streams. Algorithmica **46**(1), 97–117 (2006)
13. Y. Matias, J. S. Vitter, M. Wang, Dynamic maintenance of wavelet-based histograms, in *Proceedings of the 26th International Conference on Very Large Data Bases (VLDB 2000)* (Morgan Kaufmann, 2000), pp. 101–110
14. J. Lin, E.J. Keogh, L. Wei, S. Lonardi, Experiencing sax: a novel symbolic representation of time series. Data Min. Knowl. Disc. **15**(2), 107–144 (2007)
15. C.C. Aggarwal, J. Han, J. Wang, P.S. Yu, On clustering massive data streams: a summarization paradigm, in *Advances in Database Systems: Data Streams Models and Algorithms*, vol. 31, ed. by C.C. Aggarwal (Springer, New York, 2007), pp. 9–38
16. S. Nassar, J. Sander, Effective summarization of multi-dimensional data streams for historical stream mining, in *Proceedings of the 19th International Conference on Scientific and Statistical Database Management (SSDBM 2007)* (IEEE Computer Society, 2007), p. 30
17. X. Ma, S. Li, Q. Luo, D. Yang, S. Tang, Distributed, hierarchical clustering and summarization in sensor networks, in *Proceedings of the Joint 9th Asia-Pacific Web and 8th International Conference on Web-age Information Management and Advances in Data and Web Management (APWeb/WAIM 2007)* (Springer, 2007), pp. 168–175
18. P.P. Rodrigues, J. Gama, L.M.B. Lopes, Clustering distributed sensor data streams, in *Proceedings of the European Conference on Machine Learning and Knowledge Discovery in Databases*, vol. 5212 of LNCS, (Springer, 2008), pp. 282–297
19. M. Kontaki, A.N. Papadopoulos, Y. Manolopoulos, Continuous trend-based clustering in data streams, in *Proceedings of the 10th International Conference on Data Warehousing and Knowledge Discovery (DaWaK 2008)*, vol. 5182 of LNCS (Springer, 2008), pp. 251–262

20. A. Ciampi, A. Appice, D. Malerba, Summarization for geographically distributed data streams, in *Proceedings of the 14th International Conference on Knowledge-Based and Intelligent Information and Engineering Systems (KES 2010)*, vol. 6278 of LNCS (Springer, 2010), pp. 339–348

21. D. Malerba, A. Appice, A. Varlaro, A. Lanza, Spatial clustering of structured objects in *Proceedings of the 15th International Conference of Inductive Logic Programming (ILP 2005)*, vol. 3625 of LNCS (Springer, 2005), pp. 227–245

22. A. Ciampi, A. Appice, D. Malerba, P. Guccione, Trend cluster based compression of geographically distributed data streams, in *Proceedings of the IEEE Symposium on Computational Intelligence and Data Mining (CIDM 2011)*, part of the IEEE Symposium Series on Computational Intelligence 2011 (IEEE, 2011), pp. 168–175

23. J.G. Proakis, D.G. Manolakis, *Digital Signal Processing: Principles, Algorithms, and Applications* (Prentice-Hall, Upper Saddle River, 1996)

24. S. Mallat, *A Wavelet Tour for Signal Processing* (Academic, New York, 1998)

25. M. Garofalakis, A. Kumar, Deterministic wavelet thresholding for maximum-error metrics, in *Proceedings of the 23rd Symposium on Principles of Database Systems (PODS 2004)* (ACM, 2004), pp. 166–176

26. S. Al Wadi, M.T. Ismail, S.A.A. Karim, A comparison between Haar wavelet transform and fast fourier transform in analyzing financial time series data. Res. J. Appl. Sci. **5**(5), 352–360 (2010)

27. S. Chaudhuri, U. Dayal, An overview of data warehousing and olap technology. SIGMOD Rec. **26**(1), 65–74 (1997). ACM Special Interest Group on Management of Data, cited by (since 1996) 672

28. A. Ciampi, A. Appice, D. Malerba, A. Muolo, Space-time roll-up and drill-down into geo-trend stream cubes, in *Proceedings of the 19th International Symposium on Foundations of Intelligent Systems (ISMIS 2011)*, vol. 6804, LNCS, ed. by M. Kryszkiewicz, H. Rybinski, A. Skowron, Z.W. Ras (Springer, 2011), pp. 365–375

# Chapter 3
# Missing Sensor Data Interpolation

**Abstract** Ubiquitous sensor stations continuously measure several geophysical variables over large zones and long (potentially unbounded) periods of time. However, observations can cover neither every space location nor every time. Interpolation, i.e., the *estimation* of unknown data in each location or time of interest, can be used to supplement station records. Although in GIScience there has been a tendency to treat space and time separately, there is now great interest in analyzing data in both the domains. This suggests that integrating space and time would yield better results than treating them separately, when interpolating several geophysical fields. This chapter contributes to the investigation of spatiotemporal interpolators in a remote-sensing scenario. We describe two interpolation techniques, which use trend clusters to interpolate missing data. The former performs the estimation phase by using the *Inverse Distance Weighting* approach, while the latter uses *Kriging*. Both have been adapted to a sensor network scenario. The proposed techniques have been evaluated in a large air-climate sensor network. The empirical study compares the accuracy and efficiency of both techniques.

## 3.1 Interpolation

Interpolation is a key technique used to supplement, smooth, and standardize observational data. Historically, it has been considered a crucial task in spatial data analysis and, consequently, a wide plethora of spatial interpolation methods (deterministic, like Inverse Distance Weighting [1] and Radial Basis Functions [2], as well as stochastic, like Kriging [3]) exists in the literature. They have been largely used to recover unknown information and account for problems like missing data, energy saving, sensor default, as well as to provide support data summarization and investigation of spatial correlation between observed data [4]. More recently, ubiquity of sensing technologies has provided a huge availability of spatiotemporal data. Interpolation methods have consistently been required to handle spatiotemporal data, potentially in a streaming scenario.

A. Appice et al., *Data Mining Techniques in Sensor Networks*,
SpringerBriefs in Computer Science, DOI: 10.1007/978-1-4471-5454-9_3,
© The Author(s) 2014

### *3.1.1 Spatial Interpolators*

The spatial interpolative primitives, integrated in the majority of geographic information systems, estimate a geophysical quantity in any geographic location where the field measure is not available. The interpolated value is derived by making use of the knowledge of the nearby observed data and, sometimes, of some hypotheses or supplementary information about the data field. Inverse distance weighting (IDW) [1], radial basis functions (RBF) [2], and Kriging [3] are the most common spatial-aware techniques adopted in these cases. These techniques are specifically studied to deal with the irregular sampling of the investigated area [5, 6], or with the difficulty of describing the area by the local atlas of larger, irregular manifolds.

IDW and RBF, both deterministic interpolators, use mathematical functions to calculate an unknown field value in a geographic location, based either on the degree of similarity (IDW), or on the degree of smoothing (RBF) in relation to neighboring data points. Both methods share with Kriging (that is a statistical interpolator) the idea that the collection of field observations can be considered as the production of a correlated spatial random field with specific statistical properties. In Kriging this correlation is used to derive a second-order model of the field (the variogram). The variogram represents an approximate measure of the spatial dissimilarity of the observed data. IDW interpolation is based on a linear combination of nearby observations with weights proportional to a power of the distances. It is a heuristic but efficient approach justified by the typical power-law of the random field spatial correlation. In this sense, IDW uses the same strategy adopted by the more rigorous formulation of Kriging [7–9].

Several studies have arisen from these basic spatial interpolation approaches. In Ref. [10], the missing data of a dense network are recovered by a Kriging interpolator. By considering that the computational complexity of a variogram is cubic in the size of the observed data [11], the variogram calculus, in this study, is sped-up by processing only the areas with information holes, rather than the global data. In Ref. [12], IDW and 1-Nearest Neighbor have been used to interpolate a grid of rainfall data and re-sample data at multiple resolutions. In Ref. [13], IDW is again investigated and formulated in an adaptive way, which depends on the varying distance-decay relationship in the area under examination. The weighting parameters are varied according to the spatial pattern of the sampled points in the neighborhood. The method proves more efficient than ordinary IDW and, in several cases, also better than Kriging. These studies contribute to highlighting IDW as a deterministic, quick, and simple interpolation method, which also provides accurate interpolation results.

On the other hand, Kriging is based on the statistical properties of the random field and, hence, is expected to be more accurate regarding the general characteristics of the observations and the efficacy of the model. In any case, the accuracy of Kriging is highly dependent on a reliable estimation of the variogram [5, 14] and the variogram computation cost scales as the cube of the number of observed data [3]. This cost is prohibitive in evolving sensing environments, where statistical properties of a monitored field may change over time. In the Data Mining framework the change in the underlying properties over time is usually called concept drift [15]. It is noteworthy that the concept drift, expected in evolving data, can be a serious complication for Kriging. In fact, it may impose the repetition of costly computation of the variogram each time the statistical properties of the field change significantly. On the other hand, experimental studies reported in the literature (e.g. Ref. [13]) show that the accuracy of an IDW interpolator often approaches the accuracy of a Kriging interpolator, especially for smooth fields [16]. These considerations motivate the use of an interpolator that is accurate enough and whose learning phase can be reasonably run online with the streaming activity.

### 3.1.2 Spatiotemporal Interpolators

Recently, more and more research efforts have been made to merge traditional temporal data mining techniques with spatial interpolators. The main purpose of these studies is to transfer mature temporal data mining techniques into a joint spatiotemporal set of interpolation methods able to catch the geophysical nature of data, which are both spatially and temporally correlated. Even in the proper spatiotemporal direction, the interpolation methods are based on the idea that the sequence of observations coming from a sensor can be regarded as outcomes of a stochastic process corrupted by random noise. Hence, the model of such processes can be described (and then predicted) by means of relatively few parameters [17–19].

Initial studies have offered a partial integration of the spatial and temporal methods by first performing spatial interpolation and then reducing temporal interpolation to the application of simple methods (such as linear or spline interpolation, [20, 21]) to the sequence of snapshots of spatially interpolated data [9]. An alternative has also been explored, i.e., time series of data have been temporally interpolated for each relevant location and then used as sampled observations for the application of a traditional spatial interpolator [22].

The true integration of the spatial and temporal data component is a relatively new research field. It is essentially based on the application of a dynamic model, like the Kalman filter [23], or the Markov Random field [24], to consecutive snapshots of data. Thus, the spatial interpolation takes place according to a set of temporally changing parameters. In Ref. [25], Kriging is used for the spatial interpolation of medical images, but the statistical model of the variogram is updated according to a Kalman filter of the temporal observations. In Ref. [26], the impact of an irregular grid of sensors on data compression is analyzed and the nearest neighbor is proposed as the interpolator scheme to obtain better data compression. Instead, temporal interpolation is adopted just to assess the possible sensor clock misalignment, but no solution is formulated to provide an estimate in any spatiotemporal location of the sensed area. In Ref. [27], a methodology for the spatial and temporal interpolation of air quality data is illustrated. The methodology is two-stepped. First, non-stationary time series analysis methods are used to interpolate the data sets over periods where measurements are missing and to decompose the time series into trend and harmonic components. Then a preliminary analysis of spatial relations within the data sets and a spatiotemporal model of log-transformed data is computed. The model consists of trend and noise and represents the spatiotemporal variations in the data applied to predict the air pollution variations at unsampled points across time and space.

### 3.1.3 Challenges and New Contributions

The majority of spatiotemporal interpolation techniques described in the literature are based on the analysis of a volume of spatiotemporal data that, although big enough, are always bounded in time. In a sensing application, a (large) amount of georeferenced data arrives continuously at a high rate and is possibly subjected to data distribution drifts. In addition, the storage of this unbounded volume of geodata in a server with limited memory is, in general, subjected to a summarization process. Any future query, including interpolative requests, must operate with summaries of the geodata stream. These considerations advocate the necessity of a spatiotemporal interpolator, which accounts for both (1) the intrinsic dynamism in a geodata stream and (2) the data summarization of this large volume of data.

Therefore, in order to address both issues, we have decided to pursue the spatiotemporal strategy formulated in Ref. [9]. It advocates the importance of interpolating data by accounting for the existence of a temporal pattern in the evolution of geodata. This has paved the way for using spatiotemporal knowledge, such as trend cluster knowledge, to model geodata and processing trend clusters to achieve robust spatiotemporal interpolation functions. These functions can use either the Inverse Distance Weighting (IDW) interpolation scheme or the Kriging scheme to estimate a value at any specific location of space and point of time.

The IDW mechanism outputs a weighted average of the nearby points of the trend cluster representation for the data observed in the spatiotemporal surrounding the unknown point. The Kriging mechanism is applied to estimate unknown data, by taking into account a spatial correlation model of the sensor network. Trends are used as a guideline to transfer this model across the time horizon of the trend itself, by accounting for the dynamism of the data in the modeling phase.

### 3.2 Trend Cluster Inverse Distance Weighting

Treci (TREnd Cluster-based Inverse distance weighting) [28, 29] is a spatiotemporal interpolation technique, which operates in two phases.

The online phase (see Fig. 3.1a) consumes data snapshots as they arrive from the sensor network. It analyzes snapshots according to the count-based model of the stream, in order to determine a trend cluster segmentation of each data window. For each window, trend clusters model the spatial variation of data along the time

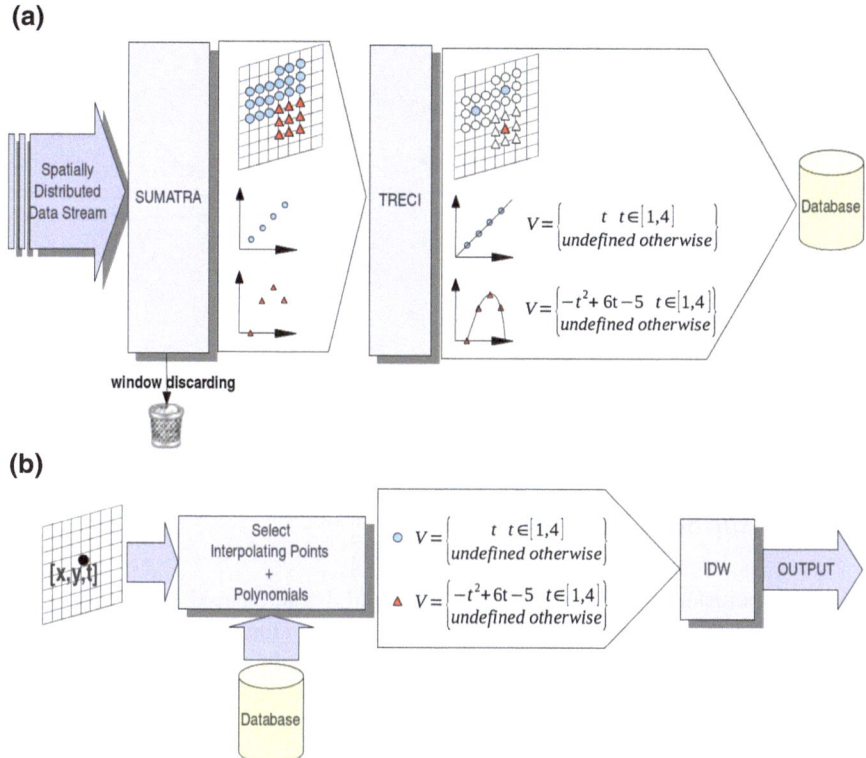

**Fig. 3.1** Treci: online summarization step and offline interpolation step. **a** Treci ONLINE, **b** Treci OFFLINE

horizon of the window. Since this model is stored in a database, while windowed data are permanently discarded, trend clusters represent the data knowledge for the future offline interpolation step. For each trend cluster, a shape-dependent sample of clustered sensors (key sensors) is extracted. The sampling algorithm is designed to keep only the information that is useful to sketch the real spatial extent of the clustered region. At the same time, a (polynomial) regression model of the time law underlying the trend time series is determined and only regression coefficients are stored in the database as a model of the trend.

The offline phase (see Fig. 3.1b), which is repeatable, retrieves the spatiotemporal knowledge surrounding the space–time point to be interpolated from the database. This knowledge is used to determine an IDW-based estimate of the field.

Details of the trend cluster discovery are already reported in Sect. 2.3, while the shape-based sampling, the polynomial interpolator learning, and the spatiotemporal IDW interpolation are described in the following subsections.

---

**Algorithm 3.1 function** sampling($\mathscr{C}, \theta) \mapsto \mathscr{S}$

---

**Require:** $\mathscr{C}$ {cluster of sensors}
**Require:** $\theta$ {density threshold}
**Ensure:** $\mathscr{S}$ {sample of key sensors extracted from $\mathscr{C}$}
1: $\mathscr{S} \leftarrow \varnothing$
2: $Q \leftarrow$ mbr($\mathscr{C}$)
3: **if** cardinality($XY, Q$)$\neq 0\%$ **then**
  4:   **if** density($\mathscr{C}, Q$)$> \theta\%$ **then**
  5:     $\mathscr{S} \leftarrow \mathscr{S} \cup \{$centroid($\mathscr{C}$)$\}$
  6:   **else**
  7:     $\mathscr{P}(\mathscr{C}) \leftarrow$subClusterQuadtree($\mathscr{C}$)
  8:     **for all** $\mathscr{C}_i \in \mathscr{P}(\mathscr{C})$ **do**
  9:       $\mathscr{S} \leftarrow \mathscr{S} \cup$ sampling($\mathscr{C}_i$)
  10:     **end for**
  11:   **end if**
12: **end if**

---

### 3.2.1 Sensor Sampling

Let $\mathscr{C}$ be a cluster of sensors. The goal is to find a shape-based sample $\mathscr{S}$ of the key sensors grouped in $\mathscr{C}$ ($\mathscr{S} \subseteq \mathscr{C}$). $\mathscr{S}$ can be stored in the database in place of $\mathscr{C}$; it represents the region covered from $\mathscr{C}$. Random sampling is the simplest way to address this task, but it poses two issues. How can we choose the number of sensors to be sampled? How can we guarantee that the randomly selected sensors maintain the information about the cluster (region) shape? To answer both questions a sampling algorithm which resorts to a *quadtree* decomposition of the clustered region can be used. The quadtree decomposition is an adaptive sampling method largely used in image processing [30, 31]. It is opportunely tailored to identify the key sensors of the cluster which are centroids in the densely populated subareas of the cluster itself. Thus, the number of sampled sensors and their location in space depend on how the cluster shape is spread across the space.

The sampling of a cluster is recursively performed according to Algorithm 3.1. First the minimum boundary rectangle $Q$ of the cluster $\mathscr{C}$ is computed (Algorithm 3.1, line 2, see Fig. 3.2a, d). The minimum bounding rectangle, also known as minimum bounding box, is the rectangle enveloping $\mathscr{C}$ that is unambiguously identified by its left inferior vertex (min(x), min(y)) and right superior vertex (max(x), max(y)) with:

$$min(x) = \min_x \{x|(x, y) \in \mathscr{C}\} \quad min(y) = \min_y \{y|(x, y) \in \mathscr{C}\}$$
$$max(x) = \max_x \{x|(x, y) \in \mathscr{C}\} \quad max(y) = \max_y \{y|(x, y) \in \mathscr{C}\}. \tag{3.1}$$

Then the density of the cluster $\mathscr{C}$ inside $Q$ (Algorithm 3.1, lines 3–4, Fig. 3.2a) is computed according to a density measure defined as follows:

$$density(\mathscr{C}, Q) = \frac{|\mathscr{C} \cap Q|}{|Q|} \times 100, \tag{3.2}$$

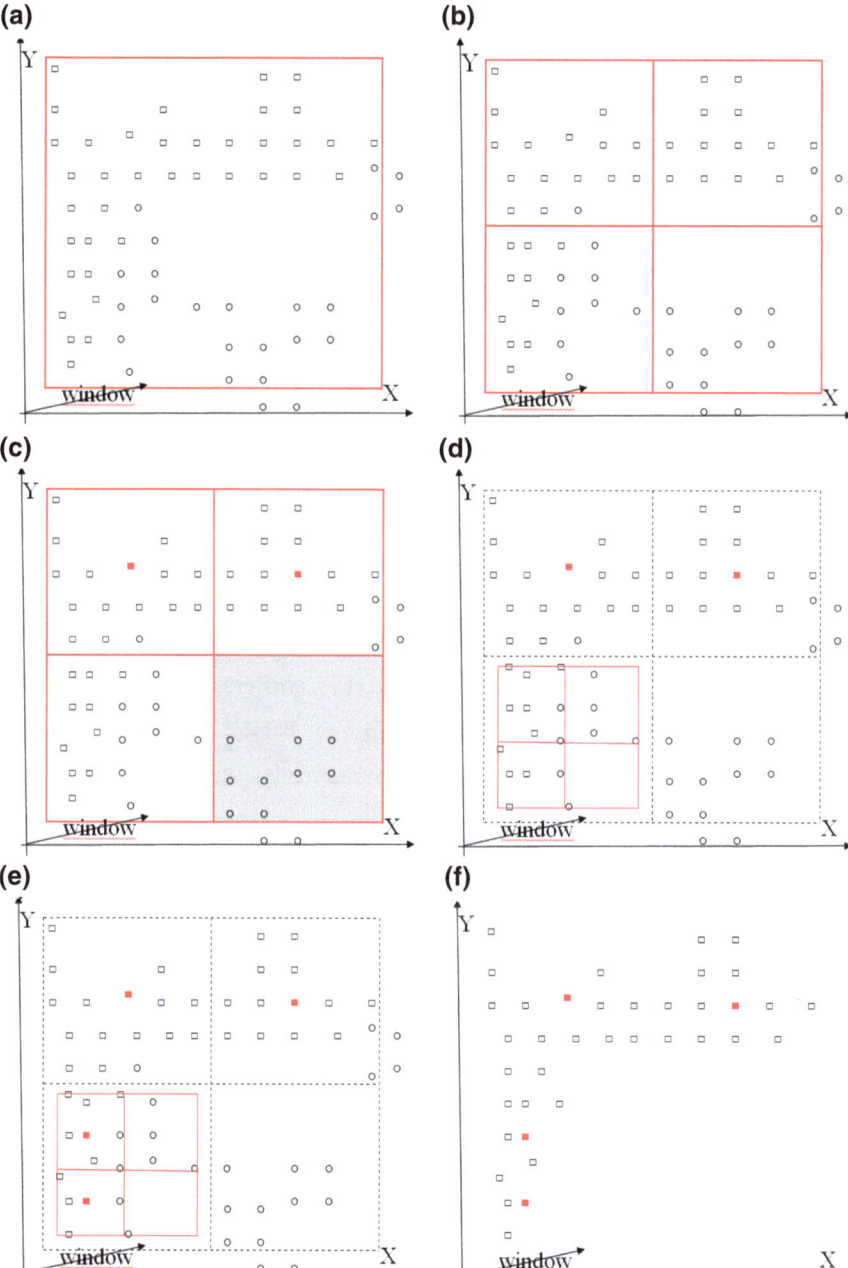

**Fig. 3.2**  An example of quadtree-based sensor sampling computed on the cluster of squares with $\theta = 75\%$. **a** MBR $density(c, Q) = 65.5\%$, **b** QuadTree decomposition, **c** sensor centroid selection, **d** recursive QuadTree decomposition of MBR rectangle, **e** recursive centroid selection, **f** sampled sensors

where $|\mathscr{C} \cap Q|$ denotes the number of sensors clustered in $\mathscr{C}$ which are spatially contained in $Q$, while $|Q|$ is the number of sensors of the network falling in $Q$. The spatial relation of containment between a 2D location $(x, y)$ and a rectangle $[(x_i, y_i), (x_s, y_s)]$ is defined as follows:

$$(x, y) \subseteq Q \Leftrightarrow x_i \le x \le x_s \wedge y_i \le y \le y_s. \tag{3.3}$$

If $density(\mathscr{C}, Q)$ is equal to zero, then $Q$ is empty and it can be discarded for the sampling (see the gray-colored quadrant in Fig. 3.2c). If $density(\mathscr{C}, Q)$ is greater than $\theta \%$ (by default $\theta = 75\%$), then $\mathscr{C} \cap Q$ can be considered a dense sub-area of $\mathscr{C}$ and its centroid node is sampled (Algorithm 3.1, lines 4–5, see red-colored sensors in Fig. 3.2c, e). Otherwise, $Q$ is decomposed into four sub quadrants (see Fig. 3.2d), that is, $Q_1$, $Q_2$, $Q_3$ and $Q_4$, and then $\mathscr{C}$ is coherently decomposed in the four subclusters falling in those quadrants, that is $\mathscr{C}_1 = c \cap Q_1$, $\mathscr{C}_2 = c \cap Q_2$, $\mathscr{C}_3 = c \cap Q_3$ and $\mathscr{C}_4 = c \cap Q_4$ (see Fig. 3.2b, d). The sampling is then recursively applied to each subcluster $\mathscr{C}_i$ (Algorithm 3.1, lines 8–10).

The quadrant decomposition of $Q$ is defined orthogonally to the axes according to $x = \frac{max(x)+min(x)}{2}$ and $y = \frac{max(y)+min(y)}{2}$, such that:

$$\begin{aligned}
Q_1 &: \left[ \left(min(x), \tfrac{max(y)+min(y)}{2}\right), \quad \left(\tfrac{max(x)+min(x)}{2}, max(y)\right) \quad \right] \\
Q_2 &: \left[ \left(\tfrac{max(x)+min(x)}{2}, \tfrac{max(y)+min(y)}{2}\right), \; (max(x), max(y)) \quad \right] \\
Q_3 &: \left[ \left(\tfrac{max(x)+min(x)}{2}, min(y)\right), \quad \left(max(x), \tfrac{max(y)+min(y)}{2}\right) \quad \right] \\
Q_4 &: \left[ (min(x), min(y)), \quad \left(\tfrac{max(x)+min(x)}{2}, \tfrac{max(y)+min(y)}{2}\right) \right].
\end{aligned} \tag{3.4}$$

The centroid of a set of sensors is computed. First, the centroid location (Fig. 3.2c) $(\widehat{x}_{\mathscr{C}}, \widehat{y}_{\mathscr{C}})$ of $\mathscr{C}$ is determined as follows:

$$\widehat{x}_{\mathscr{C}} = \frac{1}{|\mathscr{C}|} \sum_{(x,y)\in\mathscr{C}} x, \quad \widehat{y}_{\mathscr{C}} = \frac{1}{|\mathscr{C}|} \sum_{(x,y)\in\mathscr{C}} y. \tag{3.5}$$

Then the sensor of $\mathscr{C}$ which is the nearest neighbor to $(\widehat{x}_{\mathscr{C}}, \widehat{y}_{\mathscr{C}})$ is selected as the key sensor (centroid sensor) for the sampling. The centroid of $\mathscr{C}$ (see Fig. 3.2c, e) is the point location defined as follows:

$$centroid(\mathscr{C}) = \arg\min_{(x,y)\in\mathscr{C}} \{EuclideanDistance((x, y), (\widehat{x}_{\mathscr{C}}, \widehat{y}_{\mathscr{C}}))\}. \tag{3.6}$$

When no further decomposition is possible, the selected sample of sensors is output (Fig. 3.2f).

It is noteworthy that the consideration of the extracted sample of sensors in place of each original cluster drastically reduces the number of sensors processed during the interpolation phase. In this way, it speeds-up the offline interpolation phase.

---

**Algorithm 3.2 function** polynomial($\mathscr{Z}$) $\mapsto$ $poly$

---

*– Main routine*
**Require:** $\mathscr{Z}$ {the time series for the polynomial fitting}
**Ensure:** $poly$ {coefficients of the polynomial fitting $\mathscr{Z}$}
  1: $p \leftarrow$ forwardPolynomial(costantPolynomial($\mathscr{Z}$), $\mathscr{Z}$, 1)
*– forwardPolynomial($previus\,P$, $\mathscr{Z}$, deg) $\mapsto$ poly*
  1: **if** $deg \leq w - 1$ **then**
  2:   $poly \leftarrow previous\,P$
  3: **else**
  4:   $newP \leftarrow$ straightLine(residual($T^{deg}$), residual($Z$), $\mathscr{Z}$)
  5:   **if** f-test($newP$, $\mathscr{Z}$) **then**
  6:     $poly \leftarrow previuous\,P$ {The forward addition of the variable $T^{deg}$ to the polynomial is not statistically significant for the fitting of the time series $\mathscr{Z}$}
  7:   **else**
  8:     $poly \leftarrow$ forwardPolynomial($newP$, $\mathscr{Z}$, $deg + 1$)
  9:   **end if**
10: **end if**

---

This recursive subdivision algorithm has a time complexity of $O(n)$, where $n$ is the size of the cluster. It allows us to select a variable number of centroids from $\mathscr{C}$. Each centroid is strategically located in a dense area of $\mathscr{C}$, so that the necessary information to sketch the cluster shape is preserved.

## 3.2.2 Polynomial Interpolator

Let $\mathscr{Z}$ be a trend polyline prototype with length $w$. Coefficients of a polynomial interpolator, which fits points of $\mathscr{Z}$, are determined and stored in a database in place of $\mathscr{Z}$.

Let $poly : T \mapsto Z$ be the polynomial defined as follows:

$$poly(t) = \alpha + \beta_1 t + \beta_2 t^2 + \ldots + \beta_{deg} t^{deg}, \tag{3.7}$$

such that $D < w$ and the polynomial $poly(t)$ fit the series of points in $\mathscr{Z}$, according to the minimization of a cost function.

The degree $deg$ is automatically chosen ($1 \leq deg < w$) by the forward selection strategy [32], tailored for the polynomial construction. This strategy is combined with a test to estimate the ability of a polynomial to fit the time series. Once the (unknown) estimate at a generic time position $t^*$ is required, the polynomial computes:

$$Z(t^*) = poly(t^*). \tag{3.8}$$

The polynomial is built stepwise according to Algorithm 3.2. We start with $deg = 1$ and compute the (straight-line) polynomial of the variable $Z$ in the variable $T$ (see line 1 of the main routine in Algorithm 3.2). At each iteration, the ability of the current $deg$-degree polynomial (named $newP$) to fit the time series $\mathscr{Z}$ is evaluated, according to the partial F-test. The F-test [32], specifically applied in this case, allows the evaluation of the statistical significance in the improvement in the time series fitting, due to the addition of the term $t^{deg}$ to the currently constructed polynomial. If this improvement is not statistically significant, the polynomial $previousP$, that is, the polynomial previously constructed with degree $deg - 1$ (the constant polynomial if $deg = 1$) is kept and no higher degree variable is added to the final polynomial (stopping criterion, as reported in line 6 of Algorithm 3.2). On the contrary, $deg$ is incremented by one and the polynomial of degree $deg$ is forward computed by means of the straight-line regression between the residual of the dependent variable $Z$ and the residual of the $deg$-degree variable $T^{deg}$ (see line 8 of Algorithm 3.2 for the recursive call of the function $forwardPolynomial()$ and line 4 of Algorithm 3.2 for the computation of a straight line between residuals).

The residual of a variable is computed as the difference between the variable and the polynomial of degree $deg - 1$ estimating that variable. In particular, the residual of the dependent variable $Z$ is the difference between the variable itself and the current polynomial in $T$ of degree $deg - 1$ and of fitting $\mathscr{Z}$. Similarly, the residual of the independent variable $T^{deg}$ is the difference between the variable itself and the polynomial in $T$ of degree $deg - 1$, fitting the series $T^{deg}$.

The procedure is iterated until $deg = w - 1$ (see line 1 in Algorithm 3.2) or the F-test (see line 5 of Algorithm 3.2) are satisfied.

An example of the stepwise construction of a polynomial, performed according to Algorithm 3.2, is reported in Example 3.1.

**Example 3.1 (Forward selection of a polynomial)** Let us consider the case in which we intend to build the polynomial $poly$ with degree $deg = 2$,

$$poly: Z(T) = \alpha + \beta T + \gamma T^2, \tag{3.9}$$

through a sequence of parametric straight-line regressions. To this aim, we start by regressing the variable $Z$ on the 1-degree variable $T$ and building the straight line:

$$\hat{Z} = \alpha_1 + \beta_1 T. \tag{3.10}$$

The slope $\alpha_1$ and intercept $\beta_1$ are computed on the time series $\mathscr{Z}$. This equation does not fit the series exactly. By adding the 2-degree variable $T^2$, the fitting might improve. However, instead of starting from scratch and building a new polynomial with both $T$ and $T^2$, the forward strategy is exploited in the polynomial construction. First, the parametric linear polynomial is built for $T^2$, if $T$ is given, that is, $\hat{T}^2 = \alpha_2 + \beta_2 T$. Then the residuals are defined on both the independent variable $T^2$ and the dependent variable $Z$, that is:

$$\begin{aligned} T^{2'} &= T^2 - (\alpha_2 + \beta_2 T). \\ Z' &= Z - (\alpha_1 + \beta_1 T). \end{aligned} \tag{3.11}$$

Finally, the straight-line regression is determined between residuals $Z'$ and $T^{2'}$ in the time series, that is,

$$\hat{Z}' = \alpha_3 + \beta_3 T^{2'}. \tag{3.12}$$

By substituting the straight-line regressions of Eq. 3.11, the latter Equation is reformulated as follows:

$$Z - (\alpha_1 + \beta_1 T) = \alpha_3 + \beta_3 (T^2 - (\alpha_2 + \beta_2 T)). \tag{3.13}$$

This equation can be written equivalently as:

$$Z = (\alpha_3 + \alpha_1 - \alpha_2\beta_3) + (\beta_1 - \beta_2\beta_3)T + \beta_3 T^2.$$

It is proved that the polynomial reported in the last Equation coincides with the polynomial model built with $Z$, $T$ and $T^2$ (in Eq. 3.9), that is,

$$\alpha = \alpha_3 + \alpha_1 - \alpha_2\beta_3. \tag{3.14}$$
$$\beta = \beta_1 - \beta_2\beta_3. \tag{3.15}$$
$$\gamma = \beta_3. \tag{3.16}$$

A final consideration concerns the time complexity of this forward selection-based computation of a polynomial of degree $deg$, that is, $O(w \times \frac{deg(deg-1)}{2})$. This result can be interestingly combined with the consideration reported in Ref. [33] (Sect. 2.3.1, pp. 90), according to which the degree of a polynomial adequately fitting $w$ values should rarely exceed $\frac{w}{3}$.

### 3.2.3 Inverse Distance Weighting

The *Inverse Distance Weighting (IDW)* [34] is adapted in order to estimate offline the unknown value of the variable $Z$ at any space–time point $(x^*, y^*, t^*)$.

The estimate $\hat{z}(x^*, y^*, t^*)$ is computed, based on the summary $P(t_i \rightarrow t_e)$, which is the set of trend clusters stored in the database with time horizon $t_i \rightarrow t_e$, such that $t_i \leq t^* \leq t_e$. For the sake of the interpolation, we consider that, for each trend cluster, key sensors of a cluster georeference the polynomial interpolator of the trend polyline prototype associated to it. Therefore, the estimate can be computed as follows:

$$\hat{z}(x^*, y^*, t^*) = \begin{cases} poly_u(t^*) & \text{if } \exists u \in keys(P) \\ & \text{with } (x^*, y^*) \equiv (x_u, y_u), \\ \dfrac{\displaystyle\sum_{u \in keys(P)} w_{(x^*,y^*)(x_u,y_u)} \times poly_u(t^*)}{\displaystyle\sum_{u \in keys(P)} w_{(x^*,y^*)(x_u,y_u)}} & \text{otherwise,} \end{cases}$$

$$\tag{3.17}$$

where $keys(P)$ is the set of keys sampled per $P$, $poly_u(t^*)$ is the value computed at the time point $t^*$ by the polynomial interpolator, georeferenced to the key sensor $u$.

The idea behind Eq. 3.17 is that the interpolation at an unsampled point location is a function of the known values around it. In particular, it depends on them in a relation inversely proportional to the distance, i.e., the nearer a known value, the stronger its influence. According to this idea, the weights $w_{(x^*,y^*)(x_u,y_u)}$ are defined by the inverse of a power of the Euclidean distance:

$$w_{(x^*,y^*)(x_u,y_u)} = \mathbf{d}((x^*, y^*)(x_u, y_u))^{-p}. \tag{3.18}$$

Based upon Eq. 3.18, the IDW interpolation is dependent on the *power parameter* $p$, which is a positive, real number. Typically, higher values of $p$ provide more influence to the observation located closest to the unsampled position. For $p \to \infty$, IDW converges to the 1-nearest neighbor interpolation, while for $p \to 0$ it becomes an arithmetic mean. Therefore, the optimal value of this parameter is dependent on the features of the random field under study. Here, $p$ has been chosen by following the rationale in [35, 36], for which a geophysical random field has well-known self-similarity properties. Among these the correlation function of a field resembles a power-descending law of the distance $\mathbf{d}$, that is, $R_x(\mathbf{d}) \simeq |\mathbf{d}|^{-\alpha}$. By considering $p$ as the tuner of the relative influence of neighbors in the interpolation, $p = \alpha$ is chosen. On the other hand, the value of $\alpha$ is known to be related to the fractal (or Hausdorff) dimension $v$ of a field by a simple relation [35], that is, $v = n+1-\alpha/2$,, where $n$ is the field dimension. Since in the literature [36] the fractal dimension of several geophysical fields has been estimated between 2 and 3, by assuming $v = 2.5$, $\alpha = p = 3$ is achieved.

Although Eq. 3.17 considers the entire set of key sensors sampled across the networked space, it is reasonable to suppose that an *influence boundary* can be set so that the key sensors which are outside this area should not be taken at all in the computation. Thus, a spheric area is fixed around the unsampled location $(x^*, y^*)$; the key sensors contribute to the interpolation only if they are inside this spherical region. The center of the *interpolation sphere* is $(x^*, y^*)$ and the radius is a boundary parameter $b$. Based on these considerations Eq. 3.18 can be reformulated as follows:

$$w_{(x^*,y^*),(x_u,y_u)} = \begin{cases} \mathbf{d}((x^*, y^*), (x_u, y_u))^{-p} & \text{if } \mathbf{d}((x^*, y^*), (x_u, y_u)) \le b \\ 0 & \text{otherwise} \end{cases}. \tag{3.19}$$

In Treci, an automatic mechanism is used to choose $b$ at each window. This mechanism guarantees that, independently of $(x^*, y^*)$, at least one centroid is within the boundaries. The idea of automatically detecting $b$ at each window as the maximum among the distances computed between each pair of closest centroids in the set $keys(P)$ was inspired by this requirement.

The time complexity of the IDW formula linearly depends on the number of key sensors. The lower the number of sensors sampled per cluster, the faster the answer.

## 3.3 Trend Cluster Kriging

TreCK (TREnd Cluster-based Kriging) [37] is a spatiotemporal interpolation technique based on Kriging.

### 3.3.1 Basic Concepts

Kriging is a family of techniques used to interpolate the value of a variable at an unobserved location across space, starting from a known observation of its value at nearby locations and from a second-order model of the variable (*variogram*). In the original formulation of Kriging, time is ignored and any variable $Z$ is a function of the space variables only. The value at the position $(x^*, y^*)$ is estimated as follows:

$$z(x^*, y^*) = \sum_{i=1}^{N} w_i(x^*, y^*)z(x_i, y_i), \qquad (3.20)$$

where $N$ is the number of known data, called interpolation base, collected across space and each $w_i(x^*, y^*)$ is a weight to compute such a linear combination.

The weights are obtained as a solution of a system of linear equations, formulated by minimizing the variance of the prediction error. Rather than using weights based on an arbitrary function of distance, as for the Inverse Distance Weighted interpolation, the weights $w_i(x, y)$ are based on the computation of a *variogram* of the random field (details are in Ref. [5]).

A variogram is an approximate measure of statistical dissimilarity within the variable; the higher the variogram value, the more different the values assumed by the variable, on average, for that distance. Random functions, for which closely spaced values may be quite different, will have a variogram that rises quickly from the origin; random functions for which the closely spaced values are very similar will have a variogram that rises much more slowly. Given the variable $Z$, the sample variogram $\gamma(h)$ is defined as half the averaged square difference between the paired data values:

$$\gamma(h) = \frac{1}{2N(h)} \sum_{(u,v)|h_{(u,v)} \simeq h} (z(u) - z(v))^2, \qquad (3.21)$$

where $N(h)$ is the number of data pairs at a distance $h$. Eq. 3.21 assumes the existence of an isotropic model of the field [5, 34] and a proper tolerance for the distance $h$. The tolerance for the distance $h$ guarantees the consideration of an acceptable number of pairs $(i, j)$ in the empirical evaluation of the variogram.

To reduce the effect of variability, due to the unavoidable presence of noise on the measure, several research studies [3, 5, 34] have argued the appropriateness of fitting a theoretical model on the sample measure. The existing models have been inspired

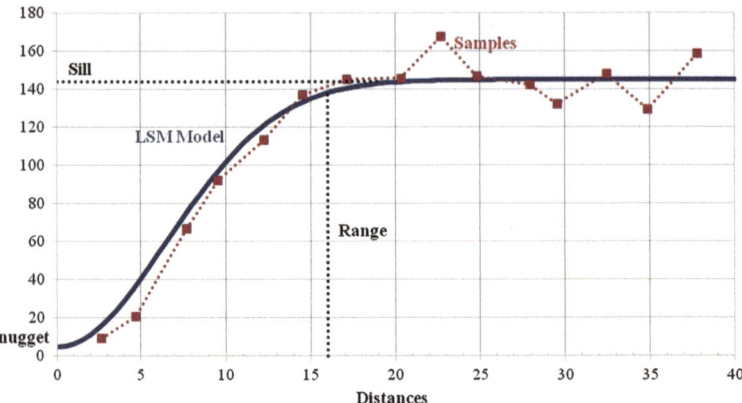

**Fig. 3.3** A sample variogram (red crosses and line) and the Gaussian model (blue line) which fits the data samples. The sill, the nugget, and the range are also highlighted

by the idea that a variogram is an approximate measure of statistical dissimilarity within the random field, so they have approximately the same behavior. They start with an initial low value (the *nugget*, that is, the value of $\gamma(h = 0)$), then increase and, after a given distance (the *range*) they approach an asymptotic value (the *sill*, which is $\gamma(h \to \infty)$). Moreover, it has been proved that the sill of a variogram is also the variance of the random field [5]. A sample variogram and a model fitted on it, using the Least Mean Square (LMS) method, is illustrated in Fig. 3.3.

### *3.3.2 Issues and Solutions*

Adapting traditional Kriging to a sensor network scenario is not a straightforward task. There are specific issues to be addressed when the variogram should be computed in a spatiotemporal setting.

#### 3.3.2.1  Spatial Non-Stationarity of Geodata

In the classical Kriging formulation, the variogram model is learned as a global function of the average squared difference in geodata, under the assumption that the function definition does not vary with space. However, we cannot suppose the spatial invariance for geodata on large extensions. Realistically, a sample variogram may vary significantly on a set of data estimated in very distant locations.

We address this issue by segmenting the surface under investigation into regions, such that the invariance of the variable statistics can be observed in each region at least up to second-order statistics like the variogram. The variogram function can be conveniently computed piecewise on these regions (called variogram regions).

### 3.3.2.2 Temporal Non-Stationarity of a Variogram

A sensor generates a time series. While a variogram represents some given spatial statistics at a specific time point, distinct variograms may arise at distinct time points. The temporal non-stationarity of a variogram can be naively faced by computing a new variogram from each data snapshot. By considering that the cost of computing a variogram scales as the cube of the number of observed data [3], this solution is not feasible in a sensor network management system, for which the assurance of a time-preserving computation is a crucial constraint.

We address this issue by defining a transfer learning technique [38], which exploits the prominent trends observed in the geodata, in order to transfer the variogram computed at a specific time point across the time horizon of the detected trends.

## 3.3.3 Spatiotemporal Kriging

The Kriging interpolation process operates in two phases. The online phase (see Fig. 3.4a) consumes snapshots as they arrive from the sensor network and pours them, window-by-window, into TreCK, which computes the interpolation base and the variogram of the window. The offline phase (see Fig. 3.4b), which is repeatable, uses the data model for the data estimation. Details on both phases are discussed in the next subsections.

### 3.3.3.1 Kriging Model Computation

Let $t_i \overset{z(T,K)}{\to} t_e$ be a data window of the $w$-sized count-based window model of the stream $z(T, K)$. The variogram $(var)$ and the interpolation base $(\mathscr{B})$ are computed for the window in three steps. Details are reported in Algorithm 3.3.

Variogram Region Segmentation [lines 1–2, Algorithm 3.3]

The data window $t_i \overset{z(T,K)}{\to} t_e$ is segmented into the set $P$ of the trend clusters, discovered with the global trend similarity threshold $\delta_G$ (line 1, Algorithm 3.3).

Let $\mathscr{P}(\mathscr{C}, \varsigma^2(\mathscr{C}))$ be the set of cluster parts collected in $P$. Each cluster $\mathscr{C} \in \mathscr{P}(\mathscr{C})$ is called the variogram region. $\varsigma^2(\mathscr{C})$ is the variance vector of the cluster $\mathscr{C}$, such that $\varsigma^2(\mathscr{C})[t]$ is the variance of values of $Z$ measured at the time $t$ from the sensors of $K$, clustered in $\mathscr{C}$ (with $t_i \leq t \leq t_e$).

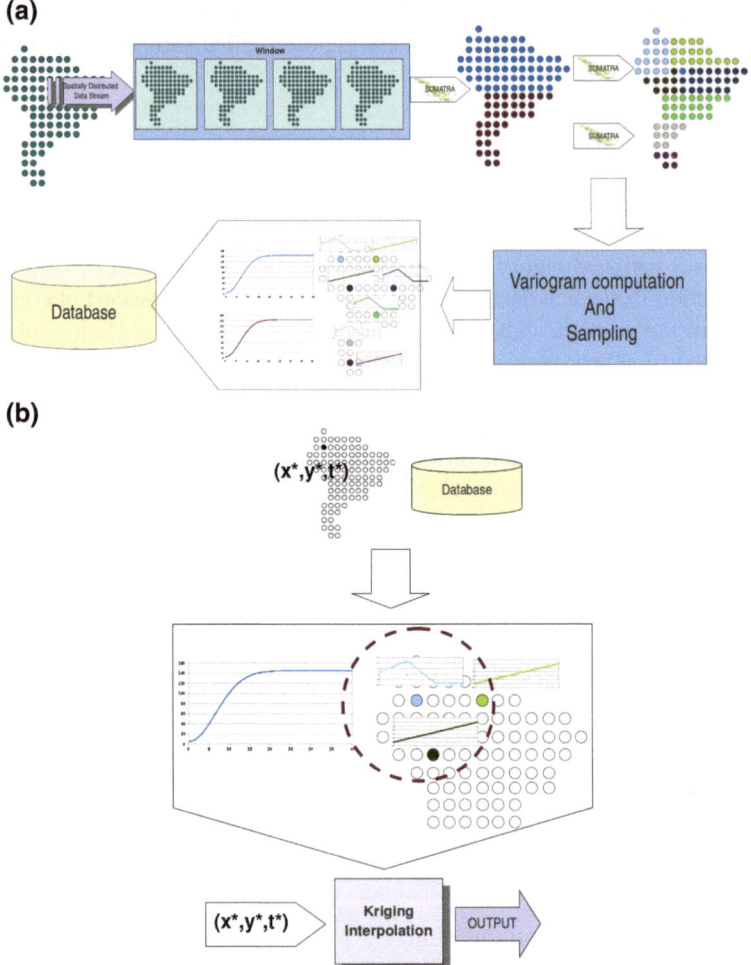

**Fig. 3.4** TreCK: online interpolation phase and offline interpolation phase. **a** TreCK ONLINE, **b** TreCK OFFLINE

$\mathscr{P}(\mathscr{C}, \varsigma^2(\mathscr{C}))$ depends on $\delta_G$. Domain knowledge expertise should aid the user in the choice of the value of $\delta_G$. This can lead to determine variogram regions, for which a global variogram can consistently fit the variability of the clustered data.

Time Series Down-Sampling [lines 3–7,11, Algorithm 3.3]

A sample of key time series $\mathscr{B}(t_i \to t_e)$ is determined. This sample is the base of the known time series, which are linearly combined during the Kriging estimation (see Eq. 3.20).

---

**Algorithm 3.3** variogram$(z(T, K), t_i \to t_e) \mapsto \mathscr{B}(t_i \to t_e), var(K, t_i \to t_e)$

---

**Require:** $t_i \overset{z(T,K)}{\to} t_e$ {a window of $w$ data snapshots measured for $Z$ from $K$ along the time horizon $t_i \to t_e$}

**Ensure:** $\mathscr{B}(t_i \to t_e)$ {Interpolation base}

**Ensure:** $var(K, t_i \to t_e)$ {Spatiotemporal variogram of $K$ with time horizon $t_i \to t_e$}

1:   $P \leftarrow$ sumatra$(t_e \overset{z(T,K)}{\to} t_e, \delta_G)$

2:   $\mathscr{P}(\mathscr{C}, \varsigma^2(\mathscr{C})) \leftarrow$ clusterSet$(P, t_i \overset{z(T,K)}{\to} t_e)$

3:   **for all** $(\mathscr{C}, \varsigma^2(\mathscr{C})) \in \mathscr{P}(\mathscr{C}, \varsigma^2(\mathscr{C}))$ **do**

4:      $P'(\mathscr{C}) \leftarrow$ sumatra$\left( \Pi_\mathscr{C} \left( t_i \overset{z(T,K)}{\to} t_e \right), \delta_C \right)$

5:      **for all** $[\mathscr{C}', \mathscr{Z}'] \in P'(\mathscr{C})$ **do**

6:         append$(\mathscr{B}(t_i \to t_e), \langle centroid(\mathscr{C}'), \mathscr{Z}' \rangle)$

7:      **end for**

8:      $\rho_\mathscr{C}, \eta_\mathscr{C} \leftarrow$ gaussianVariogram$(\mathscr{B}(t_{\frac{i+e}{2}}, \mathscr{C}))$

9:      $var(\mathscr{C}) \leftarrow (centroid(\mathscr{C}), \rho_\mathscr{C}, \eta_\mathscr{C}, \varsigma^2(\mathscr{C}))$

10:    append$(var(K, t_i \to t_e), var(\mathscr{C}))$

11: **end for**

---

In theory, this base should contain all the time series of the data window $t_i \overset{z(T,K)}{\to} t_e$ (a $w$-sized time series for each sensor of $K$). In practice, this data volume can be very large and a summary is computed. The data window can be *down-sampled*, based on the computed summary. The computed sample is that used in the variogram computation, as well as in the Kriging estimation.

The down-sampling procedure is based on the trend clusters locally discovered in a variogram region.

Let $\Pi_\mathscr{C} \left( t_i \overset{z(T,K)}{\to} t_e \right)$ be the projection of the data window, which is performed on the variogram region $\mathscr{C}$. It contains the data of the window which are measured by the sensors clustered in $\mathscr{C}$. The trend cluster set $P'(\mathscr{C})$ is discovered in $\Pi_\mathscr{C} \left( t_i \overset{z(T,K)}{\to} t_e \right)$, with local domain similarity threshold $\delta_\mathscr{C}$ (line 4, Algorithm 3.3). Each trend polyline prototype in $P'(\mathscr{C})$ is georeferenced in the centroid of the associated cluster and stored in $\mathscr{B}(t_i \to t_e)$ for the subsequent computations (lines 5–7, Algorithm 3.3).

The threshold $\delta_\mathscr{C}$ is determined with a *box plot*. Values of $\Pi_\mathscr{C} \left( t_i \overset{z(T,K)}{\to} t_e \right)$ are depicted through five summaries: the smallest observation, the lower quartile $(Q_1)$, the median $(Q_2)$, the upper quartile $(Q_3)$, and the largest observation. Given $\alpha = Q_1 - 1.5(Q_3 - Q_1)$ and $\beta = Q_3 + 1.5(Q_3 - Q_1)$, we can compute $\delta_\mathscr{C} = 0.1(\beta - \alpha)$.

Piecewise Variogram Computation and Transfer [lines 8–10, Algorithm 3.3]

The variogram is piecewise computed on the regional decomposition $\mathscr{P}(\mathscr{C}, \varsigma^2(\mathscr{C}))$, such that,

$$var(K, t_i \to t_e) = \bigcup_{(\mathscr{C}, \varsigma^2(\mathscr{C})) \in \mathscr{P}(\mathscr{C}, \varsigma^2(\mathscr{C}))} (centroid(\mathscr{C}), var(\mathscr{C})). \qquad (3.22)$$

For each variogram region, the nugget, range, and sill of the regional variogram are computed and georeferenced at the centroid of the region. We assume the Gaussian ideal variogram and use the Least Mean Square (LMS) method to fit the estimated sample variogram $\gamma_{Sample}$ with the best Gaussian model $\gamma_{Model}$. We assume that the nugget and the range are constant on the entire window. Hence, the temporal variability of the variogram is modeled by the sill. It is a vector with a distinct value for each time point between $t_i$ and $t_e$.

The shape of the Gaussian model is reported in Ref. [5] with nugget zero, range one, and sill one. The sample variogram associated to the variogram region $\mathscr{C}$ is computed by means of Eq. 3.21. It is obtained by fitting the theoretic Gaussian model on the data and represented by the nugget, range, and sill of the fitted model. Formally,

$$var(\mathscr{C}) = (\eta_{\mathscr{C}}, \rho_{\mathscr{C}}, \varsigma^2(\mathscr{C})), \tag{3.23}$$

where the range $\rho$ and the nugget $\eta$ are LMS-estimated on $\mathscr{B}(t_{\frac{i+e}{2}}, \mathscr{C})$ (i.e., the set of values of $\mathscr{B}$ timestamped at $t_{\frac{i+e}{2}}$ and georeferenced at centroids, which are sampled within $\mathscr{C}$), while the sill vector is the variance vector $\varsigma^2(\mathscr{C})$ (according to the theory reported in Ref. [5]).

The LMS-fitted model, $\gamma_{Model}(\mathscr{C}, t_{\frac{i+e}{2}})$, is found by estimating the triple, $\{\eta_{\mathscr{C}}, \rho_{\mathscr{C}}, \varsigma^2_{\mathscr{C}}\}$, such that (line 8, Algorithm 3.3):

$$\gamma_{Model}\left(\mathscr{C}, t_{\frac{i+e}{2}}\right) = \varsigma^2_{\mathscr{C}} \cdot \gamma_{Ideal}\left(h; \rho_{\mathscr{C}}\right) + \eta_{\mathscr{C}}, \tag{3.24}$$

with:

$$\{\eta_{\mathscr{C}}, \rho_{\mathscr{C}}, \varsigma^2_{\mathscr{C}}\} = \underset{\{\eta_{\mathscr{C}}, \rho_{\mathscr{C}}, \varsigma^2_{\mathscr{C}}\}}{\arg\min} \left[\left(\gamma_{Model}\left(\mathscr{C}, t_{\frac{i+e}{2}}\right) - \gamma_{Sample}\left(\mathscr{C}, t_{\frac{i+e}{2}}\right)\right)^2\right]. \tag{3.25}$$

After this estimation phase, only $\eta_{\mathscr{C}}$ and $\rho_{\mathscr{C}}$ are considered, while the sill is the variance vector of $\mathscr{C}$ (line 9, Algorithm 3.3).

### 3.3.3.2  Kriging Interpolation

Let $(x^*, y^*, t^*)$ be a space-time point, $t_i \rightarrow t_e$ be the time horizon of a window of the count-based stream model such that $t_i \leq t \leq t_e$. The estimate $\hat{z}(x^*, y^*, t^*)$ is computed according to Eq. 3.20 by linearly combining key data with weights. The key data are the data of the interpolation base $\mathscr{B}(t_i \rightarrow t_e)$, which are timestamped at $t$. The weights are the coefficients determined according to the variogram model $var(K, t_i \rightarrow t_e)$ at the time $t$.

Let $(\eta, \rho, \varsigma)$ be the nugget, range, and $t$-timestamped sill of the nearest regional variogram of $var(\mathscr{C}) \in var(K, t_i \rightarrow t_e)$. The neighboring relation is based on the computation of the Euclidean distance between $(x, y)$ and the centroid point, which georeferences the regional variogram $(\mathscr{C}) \in (K, t_i \rightarrow t_e)$ on the map. Formally,

$$(\eta, \rho, \varsigma) = (\eta_\mathscr{C}, \rho_\mathscr{C}, \varsigma^2(\mathscr{C}[t])) \text{ with } \mathscr{C} = \underset{var(\mathscr{C}) \in var(K, t_i \rightarrow t_e)}{\arg\min} d((x, y), centroid(\mathscr{C})).$$

(3.26)

In theory, the Kriging linear estimate should combine all known data of the interpolation base, but, in practice, we exploit the fact that the nearest data contribute to the unknown location more than the furthest. Based upon this consideration, we have found the estimate on just the known data falling in a neighborhood. As a neighborhood of the location $(x, y)$, we consider the sphere with center $(x, y)$ and radius $d$. This radius represents the distance, over which correlation is supposed to cut off, hence we can automatically determine $d$ by taking as distance a given percentage of the experimental range of the variogram model (by default $d = 2\rho$).

## 3.4 Empirical Evaluation

Both Treci and TreCK are written in Java. They interface the summarization system SUMATRA to discover trend clusters in a geodata stream and a database, managed by a MySQL DBMS, to store the interpolation model. The online component (summarizer) and the offline component (interpolator) of both systems are evaluated on an Intel(R) Core(TM) 2 DUO CPU $P6$1100 @2.00GHz with 3.7 GB of RAM Memory, running Ubuntu Release 12.04 (precise) $32-bit$, Kernel Linux 3.2.0-26-generic-pae.

### 3.4.1 Streams and Experimental Setup

Experiments are performed with the climatology stream *South American Climate* (details in Sect. 2.4.1.1). The computation time is used to compare the efficiency of the online systems, while the root mean squared error is used to compare the interpolation ability of the offline systems. We use both Treci and TreCK.

Experiments are run by using window size $w = 12$ (months), due to the expected yearlong periodicity of the air temperature. In Treci, the trend cluster discovery is run with a domain similarity, that is, about 10% of the field dynamics in the stream $(= 4\,°C)$. In TreCK, variogram regions are discovered with a global domain similarity $10\,°C$. This is deduced by common sense, which suggests that a spatial variation of temperature data, which differs more than $10\,°C$ across space, cannot be correctly modeled by a single variogram, provided that the dynamics of the data varies by about $40\,°C$.

The accuracy of the interpolators is evaluated by varying the percentage of sensors switched-off in the network. In this case, the online system processes only a subset of the stream, while the offline interpolator is used to estimate both available and unavailable data.

### 3.4.2 Online Analysis

We consider the entire network (all sensors are switched-on) and use both Treci and TreCK to learn the IDW/Kriging interpolation model of the stream. For Kriging, TreCK is compared to the baseline TreCK*, which learns a new piecewise regional variogram from scratch at each new snapshot of the window.

Results are reported in Table 3.1. The analysis of these results confirms that the learning process, performed by TreCK to compute a spatiotemporal variogram, obtains a remarkable reduction in the baseline computation time.

As expected, the learning process is sped-up at the expense of the interpolation accuracy. In fact, the analysis of the interpolation error reveals a result worsening in each snapshot of the stream. In any case, this is de facto a slight worsening; the average root mean squares error per snapshot grows slightly from 1.71 up to 1.94 (see Table 3.1).

On the other hand, the learning phase of TreCK is slower than the learning phase of Treci. The Inverse Distance Weighting overfits data when the interpolation model is used to estimate training data. This probably depends on the fact that the interpolation base computed with Treci is more numerous than the interpolation base computed by TreCK.

### 3.4.3 Offline Analysis

Several experimental settings are considered to analyze the accuracy of the spatiotemporal interpolator of both Treci and TreCK. In particular, this comparison is performed by varying the percentage of sensors, which are switched-off in the network, from 0 % (the network is considered in its entirety) to 10, 20, and 50 %.

**Table 3.1** Online phase (TreCK vs TreCK*, Treci vs TreCK): root mean squared error—rmse (averaged on snapshots), computation time in seconds (averaged on windows) and number of key sensors—nks (averaged on windows)

|                | Treci | TreCK | TreCK* |
|----------------|-------|-------|--------|
| Averaged time  | 21.64 | 78.5  | 311.5  |
| Averaged rmse  | 1.23  | 1.94  | 1.71   |
| Averaged nks   | 666   | 316   | 316    |

**Table 3.2** Online phase (Treci vs TreCK): rmse (averaged on snapshots)

| % | Treci | TreCK |
|---|---|---|
| 0 | 1.74 | 1.94 |
| 10 | 2.11 | 2.13 |
| 20 | 2.4 | 2.12 |
| 50 | 2.7 | 2.08 |

The average rmse per snapshot is collected in Table 3.2. This result shows the robustness of Kriging interpolation, when the network becomes more and more sparse by switching-off sensors. The Kriging interpolation model does not overfit training data, so it is more accurate than IDW when the network becomes sparse and incomplete.

This study contributes to highlighting IDW as a deterministic, quick, and simple interpolation method, which also provides accurate interpolation results. On the other hand, Kriging is based on the statistical properties of the random field and, hence, is more accurate regarding the general characteristics of the observations and the efficacy of the model. This allows us to consider Kriging (and then TreCK) to be more robust, but less efficient, for interpolating a random field of a *sparse* and *incomplete* network.

## 3.5 Summary

Trend clusters are stream patterns, which compactly represent numeric spatiotemporal data, by means of spatial clusters having prominent data trends in time. Trend cluster discovery is originally defined to summarize geodata, which are collected throughout a remote sensor network. In this chapter, we have illustrated that trend cluster discovery can be integrated as an online step in the spatiotemporal interpolation process, which permits the estimation of a variable at any location of the networked space and at any time point in the past. Both the Inverse Distance Weighting and Kriging are combined with the trend clusters to obtain spatial-aware and temporal-aware estimated values of a random field, monitored throughout a sensor network.

## References

1. D. Shepard. A Two-Dimensional Interpolation Function for Irregularly-Spaced Data, in *Proceedings of the 1968 23rd ACM National Conference, ACM '68*, New York, USA, 1968, pp. 517–524

2. G. Lin, L. Chen, A spatial interpolation method based on radial basis function networks incorporating a semivariogram model. J. Hydrol. **288**, 288–298 (2004)
3. N. Cressie, The origins of Kriging. Math. Geol. **22**(3), 239–252 (1990)
4. N. Lam, Spatial interpolation methods: a review. The Am. Cartograp. **10**, 129–149 (1983)
5. E.H. Isaaks, R. Srivastava, *An Introduction to Applied Geostatistics* (Oxford University Press, New York, 1989)
6. M. L. Stein. Interpolation of Spatial Data: Some Theory for Kriging (Springer Series in Statistics). Springer, 1 edition, June 1999
7. L. Li, P. Revesz, A comparison of spatiotemporal interpolation methods, in *GIScience*, vol. 2478, Lecture Notes in Computer Science, ed. by M. Egenhofer, D. Mark (Springer, Heidelberg , 2002), pp. 145–160
8. C. Karydas, I. Gitas, E. Koutsogiannaki, N. Lydakis-Simantiris, G. Silleos, Evaluation of spatial interpolation techniques for mapping agricultural topsoil properties in crete. in *European Association of Remote Sensing Laboratories, EARSeL eProceedings*, vol. 8, 2009, pp. 26–39
9. L. Li, X. Zhang, J. Holt, J. Tian, R. Piltner. Spatiotemporal interpolation methods for air pollution exposure. in *Proceedings of the Ninth Symposium on Abstraction, Reformulation and Approximation, SARA 2011. AAAI*, 2011, eds. by M.R. Genesereth, P.Z. Revesz
10. M. Umer, L. Kulik, E. Tanin, Spatial interpolation in wireless sensor networks: localized algorithms for variogram modeling and Kriging. Geoinformatica **14**(1), 101–134 (2010)
11. N. Cressie, *Statistics for spatial data* (Wiley, New York, 1993)
12. R.S.V. Teegavarapu, T. Meskele, C.S. Pathak, Geo-spatial grid-based transformations of precipitation estimates using spatial interpolation methods. Comput. Geosci. **40**, 28–39 (2012)
13. G.Y. Lu, D.W. Wong, An adaptive inverse-distance weighting spatial interpolation technique. J. Comput. Geosci. **34**, 1044–1055 (2008)
14. Z. Şen, A.D. Şalhn, Spatial interpolation and estimation of solar irradiation by cumulative semivariograms. Sol. Energy **71**(1), 11–21 (2001)
15. G. Widmer, M. Kubat, Learning in the presence of concept drift and hidden contexts. Mach. Learn. **23**, 69–101 (1996)
16. P. Guccione, A. Appice, A. Ciampi, D. Malerba, Trend Cluster Based Kriging Interpolation in Sensor Data Networks, in *Modeling and Mining Ubiquitous Social Media*, vol. 7472, Lecture Notes in Computer Science, ed. by M. Atzmueller, A. Chin, D. Helic, A. Hotho (Springer, Berlin, 2012), pp. 118–137
17. A.C. Harvey, *Forecasting, Structural Time Series Models and the Kalman Filters* (Cambridge University Press, Cambridge, 2001)
18. R. Shumway, D. Stoffer, *Time Series Analysis and Its Applications: With R Examples (Springer Texts in Statistics)* (Springer, New York, 2010)
19. M.M. Gaber, A. Zaslavsky, S. Krishnaswamy, Mining data streams: a review. ACM SIGMOD Record **34**(2), 18–26 (2005)
20. J.G. Proakis, D.G. Manolakis, *Digital Signal Processing: Principles, Algorithms, and Applications* (Prentice-Hall, Englewood Cliffs, 1996)
21. S.K. Mirta, J.F. Kaiser (eds.), *Handbook for Digital Signal Processing*, 1st edn. (John Wiley & Sons, Inc., New York, NY, USA, 1993)
22. L. Li, Spatiotemporal Interpolation Methods in GIS: Exploring Data for Decision Making. ETD collection for University of Nebraska - Lincoln, 2009
23. R.A. Kalman, A new approach to linear filtering and prediction problems. Trans. ASME J. Basic Eng. **82**, 35–45 (1960)
24. A. Izenman, *Modern Multivariate Statistical Techniques* (Springer, New York, 2008)
25. W.S. Kerwin, J.L. Prince, The Kriging update model and recursive space-time function estimation. IEEE Trans. Signal Process. **47**(11), 2942–2952 (1999)
26. D. Ganesan, S. Ratnasamy, H. Wang, D. Estrin, Coping with Irregular SpatioTemporal Sampling in Sensor Networks. in *2nd Workshop on Hot Topics in Networks (HotNets-II)*, 2003
27. R. Romanowicz, P. Young, P. Brown, P. Diggle, A recursive estimation approach to the spatiotemporal analysis and modelling of air quality data. Environ. Model. Softw. **21**(6), 759–769 (2006)

28. A. Ciampi, A. Appice, P. Guccione, D. Malerba, Integrating trend clusters for spatiotemporal interpolation of missing sensor data, in *Proceedings of the 11th International Symposium of Web and Wireless Geographical, Information Systems, W2GIS 2012*, 2012, pp. 203–220

29. A. Appice, A. Ciampi, D. Malerba, and P. Guccione, Using trend clusters for spatiotemporal interpolation of missing data in a sensor network. J. Spatial Info. Sci. 6(1), 119–153 (2013)

30. J. Yong, Z. Xiao-ling, S. Jun, Unsupervised classification of polarimetric SAR Image by Quad-tree Segment and SVM. in *Synthetic Aperture Radar, 2007. APSAR 2007. 1st Asian and Pacific Conference*, 2007, pp. 480–483

31. B. Kim, P. Tsiotras, Image segmentation on cell-center sampled quadtree and octree grids, in *Proceedings of the SPIE Electronic Imaging / Wavelet Applications in Industrial Processing VI*, 2009, pp. 72480L–72480L(9)

32. N.R. Draper, H. Smith, *Applied Regression Analysis* (Wiley, New York, 1982)

33. L. Fabbris, *Statistica Multivariata Analisi Esplorativa Dei Dati* (Istruzione scientifica (McGraw-Hill Companies), Collana, 1997)

34. M. Tomczak, Spatial interpolation and its uncertainty using automated anisotropic inverse distance weighting (IDW) - cross-validation/Jackknife approach. J. Geog. Info. Decis. Anal. **2**(2), 18–30 (1998)

35. S. Prigarin, K. Hahn, G. Winkler, Estimation of fractal dimension of random fields on the basis of variance analysis of increments. Numer. Anal. Appl. **4**, 71–80 (2011)

36. S. Lovejoy, D. Schertzer, *Nonlinear Dynamics in Geosciences*, Scale, Scaling and Multifractals in Geophysics: Twenty Years (Springer, New York, 2007), pp. 311–337

37. P. Guccione, A. Appice, A. Ciampi, and D. Malerba. Trend cluster based Kriging interpolation in sensor data networks. Lecture Notes in Computer Science (including subseries Lecture Notes in Artificial Intelligence and Lecture Notes in Bioinformatics), 7472 LNAI:118–137, 2012. cited By (since 1996)

38. S.J. Pan, Q. Yang, A survey on transfer learning. Knowl.Data Eng. IEEE Trans. **22**(10), 1345–1359 (2010)

# Chapter 4
# Sensor Data Surveillance

**Abstract**  A growing volume of geodata requires for appropriate data management systems, which ensure data acquisition and memory-preserving storage as well as continuous surveillance of this unbounded amount of georeferenced data. Trend cluster discovery, as a spatiotemporal aggregate operator, may play a crucial role in the surveillance process of the sensor data. We describe a computation-preserving algorithm, which employs an incremental learning strategy to continuously maintain *sliding window* trend clusters across a sensor network. The analysis of trend clusters, which are discovered at the consecutive sliding windows, is useful to look for possible changes in the data, as well as to produce forecasts of the future.

## 4.1 Data Surveillance

The widespread dissemination and the rapid increase of sensor networks, coupled with the high demand to utilize sensor data in critical real-time analysis tasks, have put the research focus on the deployment of network-integrated surveillance systems. These systems should be able to identify data that deviate from past baselines in (near) real-time. The surveillance in a sensor network comprises the steps of gathering data from one or several sensors, recognizing a behavioral pattern in data to raise an alarm in the presence of data that do not conform to the past pattern. The alarm can indicate a drift in the established behavior.

Data visualization can be a crucial means for the surveillance process. However, the visualization of sensor data cannot ignore the specific circumstance of an unbounded volume of data, which flow continuously and rapidly from geo-distributed sources. In fact, the streaming activity makes any traditional visualization of data prohibitive and ineffectual. These considerations underline the importance of extracting, in real time, a knowledge that preserves the compact representation of sensor data. This representation should highlight the prominent spatiotemporal dynamics of the data and allow us to visualize how these dynamics change.

A. Appice et al., *Data Mining Techniques in Sensor Networks*,
SpringerBriefs in Computer Science, DOI: 10.1007/978-1-4471-5454-9_4,
© The Author(s) 2014

Trend cluster discovery, as a spatiotemporal aggregate operator, may play a crucial role in the surveillance process of the sensor data applications. Initially formulated for data warehousing, trend cluster discovery gathers spatially clustered sensors, whose readings of a numeric geophysical variable show a similar trend (represented by a time series) along a time horizon. In previous chapters, we resorted to a segmentation of the time in consecutive windows, such that trend clusters can be discovered window-per-window, the representation of a trend can be compressed by applying some signal processing techniques, and trend clusters can be used to feed a trend-based cube storage of the sensor data and to interpolate data. In any case, trend clusters are always discovered along the time horizons of non-overlapping widows; in this way, trend clusters discovered in a window do not share, at least explicitly, any knowledge with trend clusters discovered in any other window.

On the other hand, the sliding window computation [1] is frequently considered in a data stream system. In this chapter, we illustrate an incremental algorithm for trend cluster discovery in sliding windows of a geodata stream. The algorithm presented seeks trend clusters in the latest data, which are constrained by a sliding window. A cluster stability index is defined. It can be computed to measure the degree of stability of trend clusters discovered throughout consecutive sliding windows and to look for drifts of data. A forecasting function can be fitted to each trend time series to produce forecasts for sensors grouped in the associated cluster.

The main challenge of discovering sliding window knowledge is how to minimize the computation cost (memory and time usage) during the discovery process. We face this challenge for trend cluster discovery by using a technique, called Sliding WIndow Trend cluster maintaining algorithm (SWIT), which efficiently maintains accurate sliding window trend clusters that arise in a sensor network. When new data are collected from the network, trend clusters are slid to fit these new data. For each trend cluster, which is currently maintained, the oldest time point is discarded. Clusters, which are spatially close and share a similar trend along the time horizon under consideration, are merged. Finally, trend clusters are spatially split when they do not fit data trends up to the present.

## 4.2 Sliding Window Trend Cluster Discovery

SWIT permits the discovery of sliding window trend clusters in a geodata stream. It uses an incremental learning strategy to slide the trend clusters of the past window, in order to fit the data which are acquired in the last round. The entire process is iterated at the acquisition of each new data snapshot and performed in (near) real time. This means that the analysis of a snapshot is completed presumably before a new snapshot is recorded. A merge-and-split procedure [2] is used.

### 4.2.1 Basics

We consider the definition of trend clusters, which is formulated in Chap. 2 (Definition 2.1).

A distance bandwidth $d$ is used to look for the *spatial closeness* relation between sensors.

**Definition 4.1** (**Spatial closeness relation between sensors**) Let $d$ be a threshold chosen for the spatial distance between sensors. The sensor $u$ is close to the sensor $v$ if $u$ is far at worst $d$ from $v$ (i.e., $distance(u, v) \le d$).

We assume that the spatial closeness relation is transitive, that is, it can be established transitively between sensors, which are related by means of other sensors that are in direct closeness one couple at a time.

A domain threshold $\delta$ is used to look for the *trend* similarity of the clustered sensors. We assume that this similarity is looked for pairwise for the sensors of a trend cluster. This similarity computation schema requires the calculus of the distance between all pairs of sensors grouped in a trend cluster.

**Definition 4.2** (**Trend similarity relation between sensors**) Let $\delta$ be the similarity threshold, $u$ and $v$ be two sensors, which measure data for $Z$ along the time horizon $H$. The trend of the sensor $u$ is similar to the trend of sensor $v$ along the time horizon $H$ if and only if,

$$\frac{1}{|H|} \sum_{t_i \in H} I(z_{t_i}(u), z_{t_i}(v)) = 0, \tag{4.1}$$

where $z_{t_i}(x)$ $(x \in \{u, v\})$ is the measure taken by the sensor $x$ at the specific time $t_i$ and $I(z_{t_i}(u), z_{t_i}(v)) = 0$ if $\|z_{t_i}(u) - z_{t_i}(v)\| \le \delta$; 1 otherwise.

The distance $\| \cdot \|$ is the absolute distance.

The trend similarity relation cannot be established transitively. In any case, Proposition 4.2.1 can be accounted for when testing the trend similarity relation in a cluster of sensors.

**Proposition 4.2.1** *Let $\mathscr{C}$ be a cluster of sensors. For each $u, v \in \mathscr{C}$, $I(z_{t_i}(u), z_{t_i}(v))$ $= 0$ if and only if $I(\arg \max_{u \in \mathscr{C}} z_{t_i}(u), \arg \min_{v \in \mathscr{C}} z_{t_i}(v)) = 0$.*

### 4.2.2 Merge Procedure

Let $w$ be the window size of the sliding window model according to the stream being processed, $P$ be the set of sliding window trend clusters maintained with the last processing round.

At the time $t_i$, the merge procedure (see Algorithm 4.1) starts after the information timestamped with the farthest time point $(t_{i-w})$ is discarded from the trend time series

**(a)**                                          **(b)**

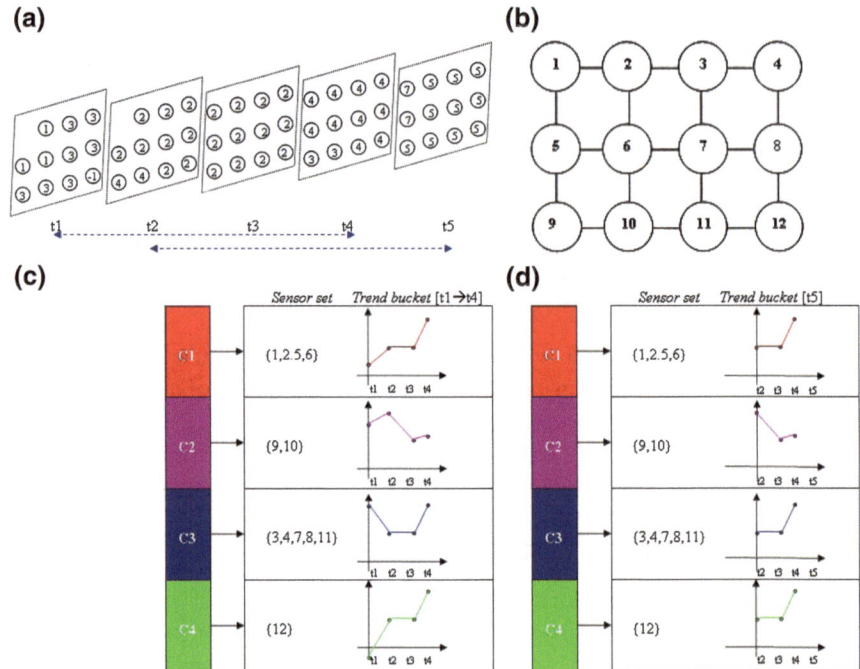

**Fig. 4.1** Sliding window process: the farthest time point ($t_1$) is discarded from each trend time series. **a** Geodata stream (with w = 4), **b** Sensor network, **c** $P(t_1 \rightarrow t_4)$, **d** $P(t_2 \rightarrow t_4)$

of each trend cluster of $P$. This happens due to the effect of the sliding window mechanism (see Fig. 4.1).

The merge procedure inputs the trend cluster set $P$, selects a random seed trend cluster $\mathcal{T} \in P$, and looks for merging trend clusters, which are close in space and similar in trend to the seed.

Let $\mathcal{T}_u = (t_{i-w+1} \rightarrow t_{i-1}, \mathcal{C}_u, \mathcal{Z}_u)$ and $\mathcal{T}_v = (t_{i-w+1} \rightarrow t_{i-1}, \mathcal{C}_v, \mathcal{Z}_v)$ be two trend clusters with the time horizon $t_{i-w+1} \rightarrow t_{i-1}$.

**Definition 4.3 (Spatial closeness relation between trend clusters)** $\mathcal{T}_u$ is close in space to $\mathcal{T}_v$ iff there exists two sensors $u \in \mathcal{C}_u$ and $v \in \mathcal{C}_v$, such that $u$ is spatially close to $v$ (according to Definition 4.2).

**Definition 4.4 (Trend similarity relation between trend clusters)** $\mathcal{T}_u$ is similar in trend to $\mathcal{T}_v$ if and only if, according to Proposition 4.2.1, for each time point $t_j$ with $t_{i-w+1} \leq t_j \leq t_{i-1}$,

$$\max\{\mathcal{Z}_u(t_j).max, \mathcal{Z}_v(t_j).max\} - \min\{\mathcal{Z}_u(t_j).min, \mathcal{Z}_v(t_j).min\} \leq \delta, \quad (4.2)$$

where *max* and *min* are aggregation statistics stored with a trend time series $\mathcal{Z}_x$ associated to the cluster $\mathcal{C}_x$ (with $x \in \{u, v\}$) in $P$.

---

**Algorithm 4.1** MergeTrendClusters(TC)

---

*– Main routine( P )*

**Require:** $P$: a set of trend clusters with time horizon $t_{i-w+1} \to t_{i-1}$
1: **for all** $\mathscr{T} \in P$ **do**
2:   $\mathscr{T} \leftarrow$ merge$(\mathscr{T}, P)$
3: **end for**
*– merge$(\mathscr{T}_u, P) \mapsto \mathscr{T}_u$*
1: **for all** $\mathscr{T}_v \in P$ **do**
2:   **if** closeInSpace$(\mathscr{T}_u, \mathscr{T}_v)$ and similarInTrend$(\mathscr{T}_u, \mathscr{T}_v)$ **then**
3:     $\mathscr{T}_u = \mu(\mathscr{T}_u, \mathscr{T}_v)$
4:     $P \leftarrow P - \{\mathscr{T}_v\}$
5:     $\mathscr{T}_u \leftarrow$ merge$(\mathscr{T}_u, P)$
6:   **end if**
7: **end for**

---

The merge operator, applied to a pair of trend clusters, computes a new trend cluster that replaces the seed of the merge process (sub-routine merge in Algorithm 4.1, lines 3–4).

**Definition 4.5 (MergeOperator $\mu$)** The operator $\mu$ inputs both $\mathscr{T}_u$ and $\mathscr{T}_v$ and computes $\mathscr{T} (= (t_{i-w+1} \to t_{i-1}, \mathscr{C}, \mathscr{L}))$, so that $\mathscr{C} = \mathscr{C}_u \cup \mathscr{C}_v$, $\mathscr{L}$ is the series of triples timestamped at the time points $t_j$ with $t_{i-w+1} \le t_j \le t_{i-1}$ and defined as follows:

$$\mathscr{L}(t_j).mean = \frac{\mathscr{L}_u(t_j).mean \times |\mathscr{C}_u| + \mathscr{L}_v(t_j).mean \times |\mathscr{C}_v|}{|\mathscr{C}_u| + |\mathscr{C}_v|}, \tag{4.3}$$

$$\mathscr{L}(t_j).min = \min\{\mathscr{L}_u(t_j).min, \mathscr{L}_v(t_j).min\}, \tag{4.4}$$

$$\mathscr{L}(t_j).max = \max\{\mathscr{L}_u(t_j).max, \mathscr{L}_v(t_j).max\}. \tag{4.5}$$

$|\cdot|$ is the cardinality of a set. The mean is computed to represent each cluster centroid in the trend cluster.

The procedure applies the merge operator to trend clusters that are close in space and similar in trend. Therefore, the output is always a "proper" trend cluster that satisfies the (transitive) spatial closeness relation (see Definition 4.1), as well as the trend cluster similarity condition (see Definition 4.2) between each pair of sensors in the output cluster.

The merge operator is recursively applied until no further merge can be performed (sub-routine merge in Algorithm 4.1, line 5) and all seeds have been considered (main routine in Algorithm 4.1, lines 1–3).

The time complexity of the procedure is $O(m^2(w-1))$ in the worst case with $m$ the number of input trend clusters.

### 4.2.3 Split Procedure

The procedure (see Algorithm 4.2) inputs the set of trend clusters from the merge procedure and the snapshot acquired in the last round.

Each input trend cluster is partitioned into sub-clusters of sensors. A sub-cluster collects data differing at worst $\delta$ from each other, at the time $t_i$ (Algorithm 4.2, lines 1–2).

The clustering is done by resorting to a contiguity-constrained clustering technique that, as pointed out in [3], permits the fitting of the requirements of learning under correlation. Clustering takes advantage of the spatial contiguity constraint between sensing devices (the one formulated in Definition 4.1) to reduce the number of possible solutions and force a fast convergence onto largely similar areal boundaries. The contiguity constraint is fulfilled by clustering sensors on a contiguity graph.

Clustering is done with a mode-seeking strategy [4], which starts from a seed sensor, to which other neighbors are added until each resulting sub-cluster ($\mathscr{C}'$) satisfies the similarity condition:

$$max \left\{ z_K(t_i)(\mathscr{C}') \right\} - min \left\{ z_K(t_i)(\mathscr{C}') \right\} \leq \delta, \tag{4.6}$$

where $z_K(t_i)(\mathscr{C}')$ is the set of measurements of $Z$ in $\langle K_{t_i}, z_K(t_i) \rangle$ for the sensors of $\mathscr{C}'$. The choice of this clustering mode is motivated by the positive properties of the seek-mode described in [5], i.e., no limit on either the geometric shape of clusters or on the number of clusters.

The cluster set $\mathscr{P}(\mathscr{C})$, that is the output of the clustering phase, is used to complete the sliding of the input trend cluster to the time $t_i$ (that of the last row). Formally, let $\mathscr{T} = (H, \mathscr{C}, \mathscr{Z})$ be the input trend cluster. For each sub-cluster $\mathscr{C}' \in \mathscr{P}(\mathscr{C})$, a trend cluster $\mathscr{T}' = (t_{i-w+1} \rightarrow t_i, \mathscr{C}', \mathscr{Z}')$ is computed for the output (Algorithm 4.2, lines 6–7), so that:

1. $\mathscr{C}'$ is the sub-cluster in $\mathscr{P}(\mathscr{C})$ (Algorithm 4.2, lines 3,6); and
2. $\mathscr{Z}'$ is the trend time series $\mathscr{Z}$, which is incremented with the statistics (minimum, maximum, mean) computed for $\mathscr{C}'$ at the time $t_i$ (Algorithm 4.2, lines 4–6).

The time complexity of the procedure is $O(n^2)$ in the worst case, with $n$ as the number of sensors spanned over the set of trend clusters.

### 4.2.4 Transient Sensors

Final notes complete the description of this process for the transient sensors, which switch their operative status from off to on and vice versa.

The former is the case of a sensor switched-on in the snapshot processed in the last round, but switched-off in the window history. This sensor is not enumerated in

---

**Algorithm 4.2** SplitTrendClusters($P, \langle K_{t_i}, z_{t_i}(K_{t_i}) \rangle) \mapsto P'$

---
**Require:** $P$: a set of trend clusters with time horizon $t_{i-w+1} \to t_{i-1}$
**Require:** $\langle K_{t_i}, z_{t_i}(K_{t_i}) \rangle$: the snapshot acquired at the last time point $t_i$
**Ensure:** $P'$: a set of slid trend clusters with time horizon $t_{i-w+1} \to t_i$
1: **for all** $\mathscr{T} \in TC$ with $\mathscr{T} = \{t_{i-w+1} \to t_i, \mathscr{C}, \mathscr{Z}\}$ **do**
2:     $\mathscr{P}(\mathscr{C}) \leftarrow \text{clustering}(z_K(t)(\mathscr{C}))$
3:     **for all** $\mathscr{C}' \in \mathscr{P}(\mathscr{C})$ **do**
4:         $\mathscr{Z}(t_i) \leftarrow \text{statistics}(z_{t_i}(\mathscr{C}'))$
5:         $\mathscr{Z}' \leftarrow \text{add}(\mathscr{Z}, \mathscr{Z}(t_i))$
6:         $\mathscr{T}' \leftarrow \text{trendCluster} \{t_{i-w+1} \to t_i, \mathscr{C}', \mathscr{Z}'\}$
7:         $P' \leftarrow \text{add}(P', \mathscr{T}')$
8:     **end for**
9: **end for**

---

any of the past trend clusters. Under the hypothesis of spatial correlation, this "new" sensor can be automatically assigned to the trend cluster that encloses the majority of its neighbors. If there is no neighbor within distance $d$, a new trend cluster is created to group the sensor and a trend time series of empty values is assigned to it. The entire window of data is acquired before this trend cluster starts to participate in both the merge and split phases of the process. During this initialization phase, the only activity is that of incrementing the trend time series with statistics measured for the sensor on the row.

The latter is the case of a sensor, enumerated in a past trend cluster, but switched-off in the row processed in the last round. One datum is expected for it. During the steady-state streaming activity, a sensor may miss a data transmission in a row without being really switched-off in the network. The sliding window phase reacts to the presence of unexpected switched-off sensors by interpolating their data (using an inverse distance-weighted sum of nearby known data [6]), putting them under surveillance and using interpolated data to complete the process of sliding the trend clusters. For each sensor, the inactivity status is declared at any missing measurement, while it is suspended at a real measurement. Sensors, kept under inactivity surveillance from the beginning of the window, are classified as switched-off, purged from the trend clusters they belong to, and no longer considered in the sliding window discovery of trend clusters.

## 4.3 Cluster Stability Analysis

A stability measure can be computed to determine if the clusters of sensors, associated to the trend clusters, are stable over consecutive sliding windows. We propose the evaluation of the stability of clustering by resorting to an error measure. It can be computed backward or forward. The higher the stability error, the greater the change in the clustering configuration.

Backward cluster stability error

This measures how much clustering changes at the time $t_i$ with respect to clustering at the past time $t_{i-1}$. It is a misclassification error percentage computed at the time $t_i$, by considering the baseline trend clusters, which are detected at the time $t_{i-1}$. Let $\mathscr{P}(\mathscr{C})$ be the clustering of sensors at the time $t_i$ and $\mathscr{P}(\mathscr{C})_B$ be the clustering at the time $t_{i-1}$, then:

$$bse(t_i) = \frac{\displaystyle\sum_{\mathscr{C} \in \mathscr{P}(\mathscr{C})} |\mathscr{C} - cluster(\mathscr{C}, \mathscr{P}(\mathscr{C})_B)|}{|K_{t_i}|}, \qquad (4.7)$$

where $cluster(\mathscr{C}, \mathscr{P}(\mathscr{C})_B) = \underset{\mathscr{C}_B \in \mathscr{P}(\mathscr{C})_B}{\arg\max} |\mathscr{C}_B \cap \mathscr{C}|$.

The backward cluster stability error can be plotted in (near) real time when a new data snapshot is processed.

Forward cluster stability error

This measures how much clustering changes at the time $t_i$ with respect to the clustering at the future time $t_{i+1}$. It is the misclassification error percentage computed at the time $t_i$, by considering the baseline trend clusters discovered at the time $t_{i+1}$. Differently from the backward cluster stability error, the forward cluster stability error is plotted with a delay in time, that is, the time required to acquire and process the next baseline snapshot.

Average cluster stability error

This is the mean of the backward cluster stability error and the forward cluster stability error.

General Considerations

We observe that if $bse(t_i) = fse(t_i)$ then $t_i$ is a stable point in the stream, otherwise $t_i$ is a drifting point in the stream. $bse(t_i) \neq 0$ indicates that a cluster merge is happening from $t_{i-1}$ to $t_i$, while $fse(t_i) \neq 0$ indicates that a cluster split is happening from $t_i$ to $t_{i+1}$.

**Example 4.3.1** Let us consider the clustering configuration associated to the sliding window trend clusters discovered with time horizon $t_{i-w+1} \rightarrow t_i$ (Fig. 4.2a) and $t_{i-w+2} \rightarrow t_{i+1}$ (Fig. 4.2b). We compute:

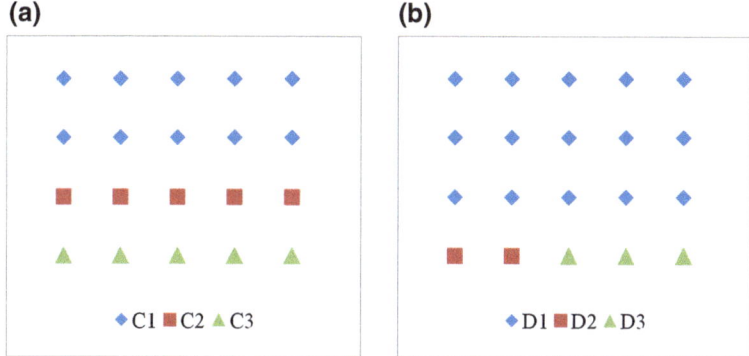

**Fig. 4.2**   Sliding window cluster configurations (from $t_i$ to $t_{i+1}$): $C_1$ and $C_2$ are merged into $D_1$, while $C_3$ is split in $D_2$ and $D_3$. **a** Clustering ($t_{i-w+1} \rightarrow t_i$), **b** Clustering ($t_{i-w+1} \rightarrow t_{i+1}$)

$$bse(t_{i+1}) = \frac{|D_1 - C_1| + |D_2 - C_2| + |D_3 - C_2|}{20} = \frac{5 + 0 + 0}{20} = 0.25 \geq 0,$$

$$fse(t_i) = \frac{|C_1 - D_1| + |C_2 - D_1| + |C_3 - D_3|}{20} = \frac{2 + 0 + 0}{20} = 0.1 \geq 0,$$

which correctly give warning of merge of clusters $C_1$ and $C_2$ into $D_1$ and the split of the cluster $C_3$ in $D_2$ and $D_3$. The merge event means that sensors having distinct trends until $t_i$ start to have the same trend from $t_{i+1}$. The split event means that sensors having the same trend until $t_i$ start to have a distinct trend from $t_{i+1}$.

## 4.4 Trend Forecasting Analysis

The time series forecasting theory is concerned with fitting a function of time on the data of a time series. Therefore, a forecasting function can be fitted to the trend time series of a trend cluster and used to extrapolate data prediction of clustered sensors for the future. The forecasting error can be analyzed. The higher the error means, the more the real data differ from the forecasts produced. This may indicate a drift in the expected trend model according to the data, which are produced.

Several forecasting models are described in the literature; a plethora of these models broadly uses the linear predictive theory and builds the model as a linear combination of recent data. We consider the exponential smoothing procedures [7], which provide forecasting functions accurate enough and easy to increment when a new data measurement is available in the time series, according to the sliding window model.

## 4.4.1 Exponential Smoothing Theory

In the exponential smoothing theory, a time series is taken as the sum of a trend, a seasonal (periodic) component, and a residual fluctuation. The trend is the average progress of the variable in time and is usually described by a simple and slowly varying function (e.g. linear). The seasonality is an emerging (close to-) periodic behavior in the series, once the general trend is removed from data. The residual fluctuation (error $\epsilon$) is any random unexplained variation in data. It is usually supposed to have little extent with respect to trend and seasonality and to be distributed as white noise (i.e., an uncorrelated Gaussian zero-mean process).

Let $z(t)$ be a time series of numerical measures for $Z$ which are routinely sampled over a time interval. A forecasting model can be produced by composing the trend ($\tau$, the *intercept* and $\iota$, the *slope*) and the seasonality ($\nu$) contributions predicted from past samples. The usual way to combine these components is by the additive model:[1]

$$z(t) = [\tau(t) + \iota(t) + \nu(t)] + \epsilon(t). \tag{4.8}$$

To determine $\tau(\cdot)$, $\iota(\cdot)$ and $\nu(\cdot)$, i.e., the predictable part of the model, the exponential smoothing models of Brown, Holt, and Winters are revised here. Brown's model and Holt's model perform ad hoc procedures that account for the trend in a time series, but they neglect seasonality. Winters' model accounts for both trend and seasonality. None of these models can be considered the best in absolute. For this reason, the best model should be chosen from time to time, according to the properties of the data.

### 4.4.1.1 Brown's Model

Brown's model averages past data collected in the time series up to the current sample. The recent data are given more weight than the older ones. To obtain a fading weight schema [7], the trend component is defined in an exponential recursive formula

$$\tau(t_i) = \begin{cases} z(t_1) & i = 1 \\ \alpha z(t_i) + (1 - \alpha)\tau(t_{i-1}) & otherwise \end{cases}. \tag{4.9}$$

The starting step (in $t_1$) is defined in [8]. The recursive step starts from $t_2$. The coefficient $\alpha \in [0, 1]$ balances the importance of the datum in $t_i$ with respect to the past; past data are recursively accumulated in $\tau(t_{i-1})$. The value forecast $\hat{z}$ for the time $t_{i+1}$ is $\hat{z}(t_{i+1}) = \tau(t_i)$.

---

[1] An alternative, rarely used, can be the multiplicative model.

#### 4.4.1.2 Holt's Model

Holt's model improves the previous one by correcting a linear tendency in the trend part [7], such that

$$\tau(t_i) = \begin{cases} z(t_i) & i = 1, 2 \\ \alpha z(t_i) + (1 - \alpha)(\tau(t_{i-1}) + \iota(t_{i-1})) & otherwise \end{cases}, \quad (4.10)$$

$$\text{with } \iota(t_i) = \begin{cases} \tau(t_2) - \tau(t_1) & i = 1 \\ \beta(\tau(t_i) - \tau(t_{i-1})) + (1 - \beta)\iota(t_{i-1}) & otherwise \end{cases}. \quad (4.11)$$

The starting step is defined in [8]. The recursive step starts from $t_3$. The coefficient $\beta \in [0, 1]$ weights a linear trend in the prediction model. The value forecast is $\hat{z}(t_{i+1}) = \tau(t_i) + \iota(t_i)$.

#### 4.4.1.3 Winters' Model

Winters' model finally assumes the existence of seasonality with a period $w$ in the time series. The recursive relations for trend, in bias and linear tendency, and seasonality [7] are defined as follows:

$$\begin{aligned} \tau(t_i) &= \alpha(z(t_i) - \nu(t_{i-w})) + (1 - \alpha)(\tau(t_{i-1}) + \iota(t_{i-1})), \\ \iota(t_i) &= \beta(\tau(t_i) - \tau(t_{i-1})) + (1 - \beta)\iota(t_{i-1}), \\ \nu(t_i) &= \gamma(z(t_i) - \tau(t_i)) + (1 - \gamma)\nu(t_{i-w}), \end{aligned} \quad (4.12)$$

with $t_i > t_w$; the coefficient $\gamma \in [0, 1]$ balances for the seasonal component $\nu$.

The forecasting process starts at time $t_{w+1}$, while the procedure, described in [8], requires $L$ ($L > 1$) initial periods to start the iterations. The value forecast is $\hat{z}(t_{i+1}) = \tau(t_i) + \iota(t_i) + \nu(t_i)$.

### 4.4.2 Trend Cluster Forecasting Model Update

For each trend cluster, the coefficients of the associated forecasting model, $\mathcal{M}$, are updated, in order to estimate the next point expected in the trend time series. This estimation process is pursued on the mean statistic, so that this estimated value can be used to forecast the next measurements of sensors enumerated in the associated cluster. The forecast value is computed according to the recursive formulation of the exponential smoothing models reported in Sect. 4.4.1. The estimate is carried out by considering Brown, Holt, and Winters' models. By accounting for the formulation in Sect. 4.4.1, the prediction for the next time is then computed after that:

1. $\tau(t_i)$ is computed for Brown's model,
2. $\tau(t_i)$ and $\iota(t_i)$ are computed for Holt's model,
3. $\tau(t_i)$, $\iota(t_i)$ and $\nu(t_i)$ are computed for Winters' model.

In case (1), $\tau(t_i)$ is computed recursively from $\tau(t_{i-1})$ (see Eq. 4.9). In case (2), $\tau(t_i)$ and $\iota(t_i)$ are computed recursively from $\tau(t_{i-1})$ and $\iota(t_{i-1})$ (see Eq. 4.11). In case (3), $\tau(t_i)$, $\iota(t_i)$ and $\nu(t_i)$ are computed recursively from $\tau(t_{i-1})$, $\iota(t_{i-1})$ and $\nu(t_{i-\omega})$ (see Eq. 4.12). For the recursive computation, these coefficients are kept in the forecasting model $\mathcal{M}$, associated to the trend cluster. When a trend cluster is computed by means of a merge operation, the coefficients of the original models are combined in a merged model by a weighted average. The weights are proportional to the size of original clusters.

## 4.5 Empirical Evaluation

SWIT is written in Java. It is evaluated on an Intel(R) Core(TM) 2 DUO CPU $P61100$ @2.00GHz with 3.7 GB of RAM Memory, running Ubuntu Release 12.04 (precise) $32 - bit$, Kernel Linux 3.2.0–26-generic-pae.

### 4.5.1 Streams and Experimental Goals

Experiments are performed with the temperature stream of *Intel Berkeley Lab* and the climatology stream of *South American Climate* (details in Sect. 2.4.1.1).
    For both streams, we study:

1. the accuracy and efficiency of the incremental discovery of sliding window trend clusters;
2. the stability of clustering throughout the sliding windows of the stream;
3. the forecasting ability of the sliding window trends.

### 4.5.2 Sliding Window Trend Cluster Discovery

SWIT is compared to the baseline system, denoted W-by-W, which discovers trend clusters by processing *all* the data of each sliding window. W-by-W performs trend cluster discovery by integrating Algorithms 2.1, 2.2 (see their description in Chap. 2) in a sliding window framework. Purity of trend clusters is evaluated by resorting to the same similarity schema of SWIT (see Definition 4.2).
    The comparison is based on the analysis of the computation time, summarization error, and clustering error. The computation time (in millisecs) measures the time spent to complete the discovery of trend clusters when a new snapshot is acquired in the stream.

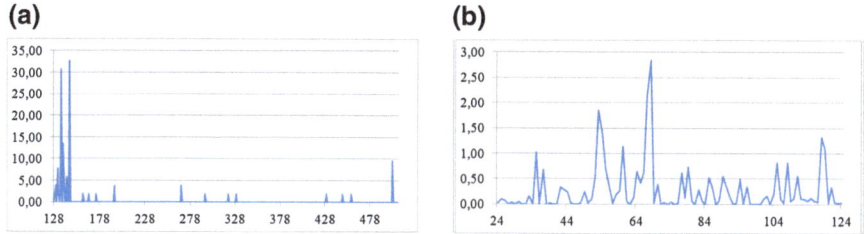

**Fig. 4.3**  SWIT: average stability error (in percentage) is plotted per sliding window. **a** IBL ($w = 128$ and $\delta = 2.5^oC$), **b** SAC ($w = 24$ and $\delta = 5^oC$ )

The summarization error is the root mean square error computed when the data of a sensor are summarized by the trend cluster.

The clustering error measures the percentage of sensors which are grouped in an unexpected cluster. The expected clusters are the trend clusters discovered by W-by-W. For each trend cluster $\mathscr{T}$ discovered by SWIT, we determine the baseline trend cluster $\mathscr{B}$ discovered by W-by-W, which groups the majority of its sensors. Sensors of $\mathscr{T}.\mathscr{C} \cap \mathscr{B}.\mathscr{C}$ are correctly classified; sensors of $\mathscr{T}.\mathscr{C} - \mathscr{B}.\mathscr{C}$ are misclassified. The clustering error is the percentage of misclassified sensors.

Both SWIT and W-by-W are run by varying the window size $w$ and the domain similarity threshold $\delta$. Results are collected in Table 4.1. The analysis of results leads to several considerations.

The incremental learning strategy speeds-up the discovery process. The reduction of the amount of data to be processed (a single snapshot rather than the window) saves computation time as a consequence. The computation time of SWIT is always much lower than the computation time of W-by-W, independently of $w$ and $\delta$.

The accuracy of the trend clusters is not decreased when they are discovered by SWIT. In IBL, the summarization errors of both systems are comparable. In SAC, the summarization error of SWIT is even lower than the summarization error of W-by-W.

The clustering error is, in general, low (with peaks of 8 % in IBL and 26 % in SAC). This means that the incremental startegy provides a good approxiamtion of the expected clusters. The error tends to decrease when enlarging $w$.

### 4.5.3  Clustering Stability

The cluster stability error is studied for trend clusters maintained by SWIT. We consider trend clusters discovered with $w = 128$ and $\delta = 2.5^oC$ for IBL, $w = 24$ and $\delta = 4^oC$ for SAC. The average stability error (in percentage) is plotted in Fig. 4.3a for IBL and Fig. 4.3b for SAC. A peak of high stability error reveals a time point, where the clustering of sensors changes with the time. The analysis of cluster stability provides interesting insights into the nature of these data.

**Table 4.1** SWIT versus W-by-W (Intel Berkeley Lab and South American Air Climate): statistics computed per sliding window are averaged on the number of sliding windows

| w | $\delta$ | SWIT | | | W-by-W | | | clustering error% |
|---|---|---|---|---|---|---|---|---|
| | | time (ms) | rmse | #trend clusters | time (ms) | rmse | #trend clusters | |
| *Intel Berkeley Lab* | | | | | | | | |
| 32 | 1.25 | 2.06 | 0.46 | 9.14 | 15.99 | 0.46 | 9.00 | 8.09 |
| 32 | 2.5 | 2.38 | 0.75 | 3.53 | 24.31 | 0.74 | 3.57 | 9.67 |
| 32 | 5 | 1.76 | 0.82 | 1.70 | 27.72 | 0.82 | 1.44 | 0.02 |
| 64 | 1.25 | 4.74 | 0.43 | 11.23 | 29.47 | 0.43 | 10.20 | 6.30 |
| 64 | 2.5 | 4.01 | 0.73 | 4.68 | 49.07 | 0.73 | 4.41 | 8.24 |
| 64 | 5 | 2.32 | 0.81 | 2.04 | 61.49 | 0.81 | 1.62 | 0.02 |
| 128 | 1.25 | 17.14 | 0.39 | 12.47 | 65.56 | 0.40 | 12.54 | 5.49 |
| 128 | 2.5 | 11.48 | 0.70 | 7.22 | 104.89 | 0.70 | 5.46 | 4.46 |
| 128 | 5 | 3.59 | 0.80 | 2.39 | 134.37 | 0.79 | 2.05 | 0.02 |
| *South American Air Climate* | | | | | | | | |
| 6 | 2 | 6069.00 | 0.76 | 188.64 | 18876.46 | 0.81 | 106.74 | 23.83 |
| 6 | 4 | 9595.95 | 1.43 | 56.08 | 44017.49 | 1.53 | 36.31 | 16.00 |
| 6 | 8 | 20689.79 | 3.06 | 14.15 | 90789.98 | 3.30 | 11.62 | 4.88 |
| 12 | 2 | 5730.96 | 0.72 | 232.00 | 25279.87 | 0.77 | 133.39 | 26.93 |
| 12 | 4 | 9024.30 | 1.34 | 70.64 | 60952.52 | 1.45 | 46.72 | 17.50 |
| 12 | 8 | 16766.44 | 2.87 | 17.89 | 137296.9 | 3.09 | 14.65 | 5.93 |
| 24 | 2 | 5666.98 | 0.69 | 279.00 | 36863.70 | 0.74 | 155.93 | 26.99 |
| 24 | 4 | 8456.80 | 1.35 | 96.60 | 95688.27 | 1.38 | 49.97 | 19.91 |
| 24 | 8 | 14791.74 | 2.82 | 21.19 | 234658.70 | 2.94 | 16.31 | 6.98 |

The outdoor temperature values, which are collected in SAC, form a (near) stable stream. The low stability error denotes a slight modification of clustering, which affects 3 % of sensors on average.

The indoor temperature values, which are collected in IBL, form a stream which is characterized by long periods of stable clusters (where the stability error is 0), which are interrupted by some *pervasive* change in the cluster configuration. In these cases, the cluster stability error grows up to 30 %. This is consistent with the description of these data; they are noised data, subject to sudden changes.

## *4.5.4 Trend Forecasting Ability*

The forecasting error is studied for the forecasting models maintained on sliding windows trends. We consider both Brown's model, Holt's model and Winters' model whose coefficients are incremented over the maintained trends. The forecasting model of a trend cluster is used to produce forecasts of the next snapshot for sensors grouped in the cluster. The forecasting error is plotted per snapshot in Fig. 4.4 for IBL and Fig. 4.5 for SAC. The higher the error, the greater the difference between the real values and the produced forecasts.

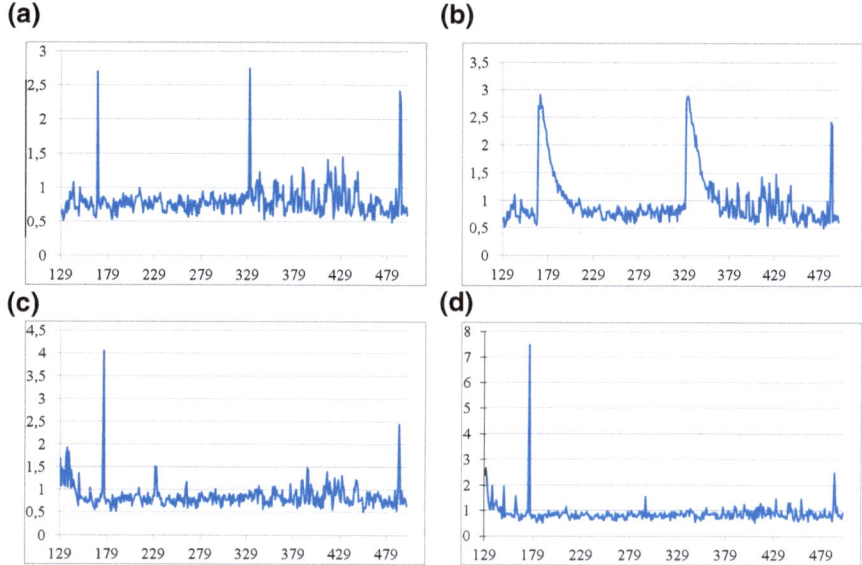

**Fig. 4.4** Forecasting error (IBL): root mean square error plotted per snapshot. **a** Brown model, **b** Holt model, **c** Winters model $w = 32$, **d** Winters model $w = 64$

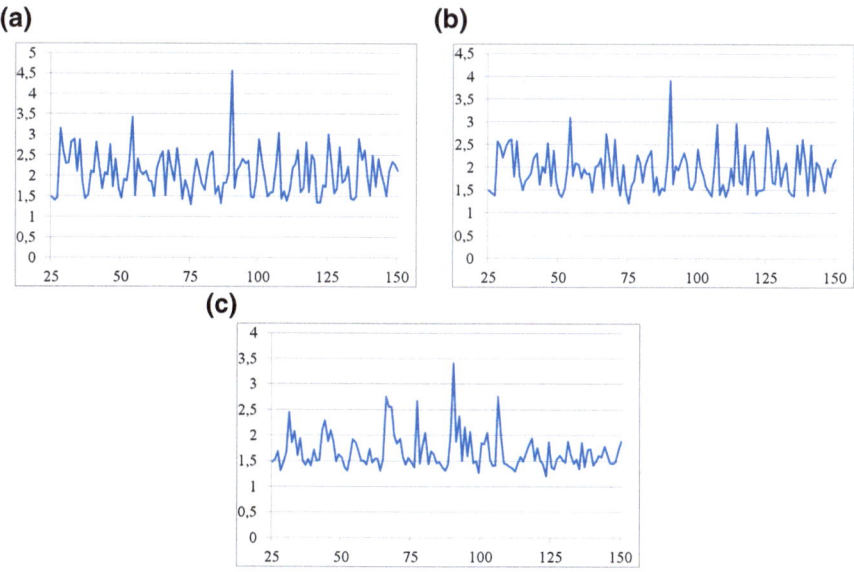

**Fig. 4.5** Forecasting error (SAC): root mean square error plotted per snapshot. **a** Brown model, **b** Holt model, **c** Winters model $w = 12$

Forecasting models have different performances. Brown's model is the most accurate for IBL, where no periodicity is known for data. Winters' model is the most accurate for SAC, where the year-long periodicity of air temperature can be used to appropriately set up the period $\omega$.

In any case, the forecasting errors of the three models exhibit peaks, which are, in general, in the same time periods. A high forecasting error indicates a time period where the data drift from the behavior established by the forecasting model. For IBL, the three models detect a drift between time $t_{168}$ and time $t_{175}$, as well as a drift between time $t_{492}$ and time $t_{493}$. For SAC, a drift is observed in all three models at the time $t_{90}$.

## 4.6 Summary

In this chapter we have illustrated an incremental strategy to maintain sliding window trend clusters in a sensor network. A forecasting function can be fitted to each trend time series to produce forecasts for sensors grouped in the associated cluster. The proposed strategy reduces the amount of processed data by saving computation time. The stability of clusters discovered throughout consecutive sliding windows can be monitored to look for drifts of data. The forecasting error can be used to observe drifts when data differ greatly from the forecasts produced.

## References

1. M.M. Gaber, A. Zaslavsky, S. Krishnaswamy, Mining data streams: a review. ACM SIGMOD Rec. **34**(2), 18–26 (2005)
2. A. Appice, D. Malerba, A. Ciampi, in *Continuously Mining Sliding Window Trend Clusters in a Sensor Network*, ed. by S.W. Liddle, K.-D. Schewe, A.M. Tjoa, X. Zhou. 23rd International Conference on Database and Expert Systems Applications, DEXA 2012, Proceedings, Part II. Lecture Notes in Computer Science, vol. 7447 (Springer, 2012), pp. 248–255
3. P. Legendre, in *Constrained Clustering* ed. by P. Legendre, L.L. Developments in Numerical Ecology (Springer-Verlag, New York, 1987), pp. 289–307
4. J. Kittler, A local sensitive method for clustering analysis. Pattern Recognit **8**, 22–33 (1976)
5. B. Thiesson, J. Kin, Fast variational mode-seeking, in *Proceedings of the 15th International Conference on Artificial Intelligence and Statistics, AISTATS 2012*
6. M. Tomczak, Spatial interpolation and its uncertainty using automated anisotropic inverse distance weighting (IDW) - cross-validation/Jackknife approach. J. Geog. Inf. Deci. Anal. **2**(2), 18–30 (1998)
7. A.C. Harvey, *Forecasting, Structural Time Series Models and the Kalman Filters* (Cambridge University Press, Cambridge, 2001)
8. C. Vercellis, *Business Intelligence: Data Mining and Optimization for Decision Making* (Wiley Press, New Jersey, 2009)

# Chapter 5
# Sensor Data Analysis Applications

**Abstract** A PhotoVoltaic (PV) plant is a power station which converts sunlight energy into electric energy. In the last decade, PV plants have become ubiquitous in several countries of the European Union, due to a valuable policy of economic incentives (e.g., feed-in tariffs). Today, this ubiquity of PV plants has paved the way to the marketing of new smart systems, designed to monitor the energy production of a PV plant grid and supply intelligent services for customer and production applications. In this chapter, we start moving in this direction by fulfilling the urgent request of PV customers and PV companies to enjoy knowledge-based managing and monitoring services, integrated within a PV plant network. In particular, we illustrate a business intelligence solution developed to monitor the efficiency of the energy production of PV plants and a data mining solution for the fault diagnosis in PV plants.

## 5.1 Monitoring Efficiency of PV Plants: A Business Intelligence Solution

Monitoring performances of solar PV plants has become extremely important due to the high cost of maintenance operations, the reduction of incomes due to unexpected faults and the degradation of non-monitored performances. It becomes essential to ensure high performance, low downtime, and automatic fault detection. In this Section, we present a remote distributed system, called *Sun Inspector*, which permits the monitoring of the efficiency of the energy production of PV plants. Sun Inspector, developed according to the "Software as Service" (SaaS) paradigm, offers a well designed monitoring and analytical system, which assists installers and owners in reducing the cost of efficiency monitoring and plant maintenance.

Sun Inspector includes services for data collection, summarization (based on trend cluster discovery), synthetic data generation, supervisory monitoring, model learning, and visualization. On-site weather data, as well as energy production data, which are measured by panel strings, inverters, and transformers, are continuously

A. Appice et al., *Data Mining Techniques in Sensor Networks*,
SpringerBriefs in Computer Science, DOI: 10.1007/978-1-4471-5454-9_5,
© The Author(s) 2014

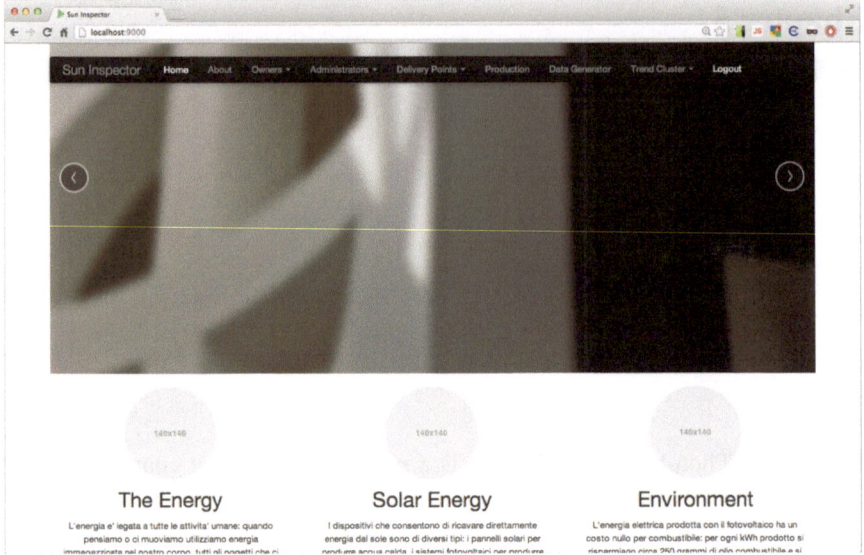

**Fig. 5.1**  Sun Inspector web interface

collected by Sun Inspector. Raw data, initially stored in a relational database, are summarized using the trend cluster discovery and saved in a data warehouse for any future queries and analysis. To simulate PV plan productions, synthetic data can be produced and plotted on charts. Finally, end users can query and display both energy production data and trend clusters, in order to monitor performances and view anomalies.

All these services can be used through a web interface. Figure 5.1 shows the main page of Sun Inspector. Using the navigation bar, it is possible to access all the services. A demo of Sun Inspector is available at http://www.kdde.uniba.it/suninspector, while the source code is available at the project code repository.[1]

## 5.1.1 Sun Inspector Architecture

The architecture of the proposed system consists of several components, each of which is in charge of performing specific tasks as shown in Fig. 5.2. A brief description of each component is given below.

---

[1] http://bitbucket.org/kddeuniba/suninspector

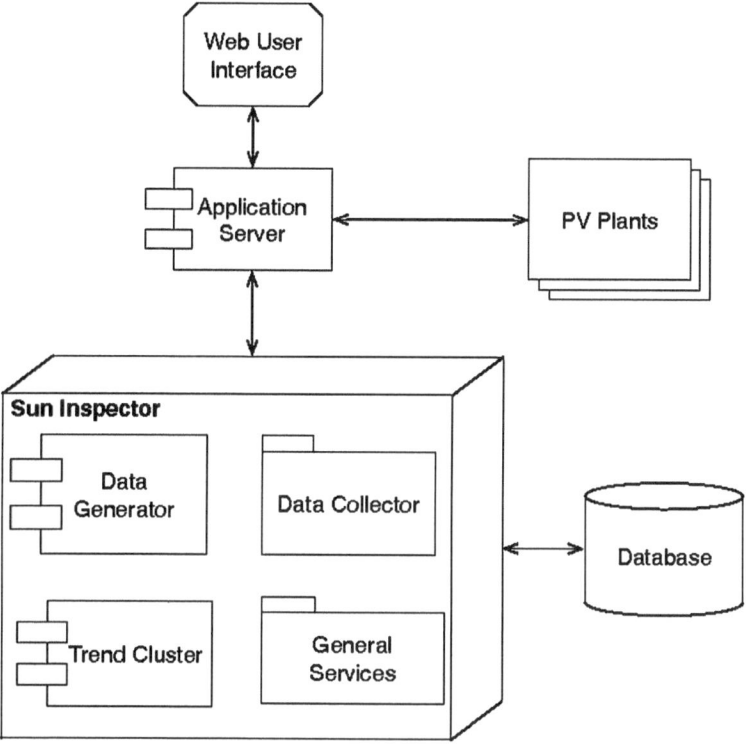

**Fig. 5.2**  Sun Inspector system architecture

### 5.1.1.1  General Services

This component allows PV companies and PV owners, to register a PV plant to obtain information about a PV plant, to display energy production data, and to access the Business Intelligence services. Figure 5.3 displays an example of energy production report generated by Sun Inspector. By selecting a PV plant from the bottom table and a date under analysis, Sun Inspector generates an energy production bar chart. Using these reports end users can monitor PV plant performances and check for daily production anomalies.

The General Services Component is in charge of administering the database and providing authorized access to the saved data. It enables the execution of all the services scheduled by the Sun Inspector administrators through the web interface. Figure 5.4 displays the web page to save a new PV plant in Sun Inspector.

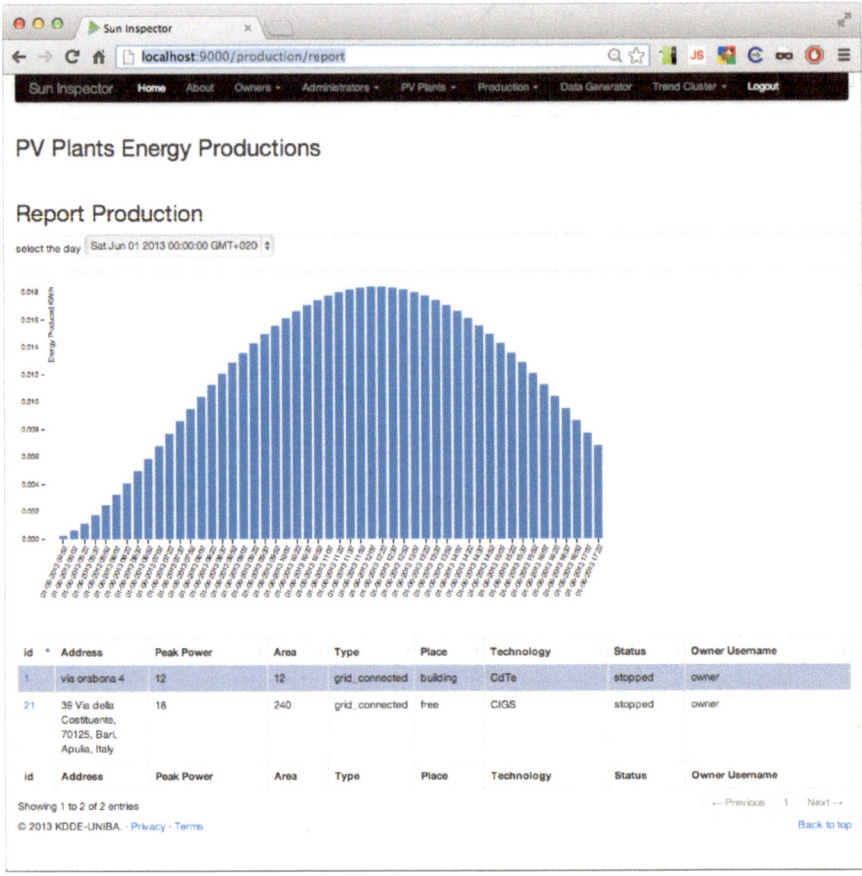

**Fig. 5.3** Sun Inspector web interface: the web page to view PV plant energy production reports

### 5.1.1.2  Data Collector

The data collector component allows Sun Inspector to obtain data from PV plants, which are registered in Sun Inspector. All the production data are offered via a REST Web Service [1], which accepts the data formatted as tab-separated values.

In order to use the data collection service, data loggers or micro-controllers measure and acquire the signals from the PV plants and transmit them to Sun Inspector through the web. Each transmission contains the identifier of the PV plant stored in Sun Inspector, the timestamp, and the measures of the energy production and additional parameters. After receiving these data, Sun Inspector stores them in the database. Figure 5.5 displays an example of energy production data saved by the data collector component and visualized through Sun Inspector.

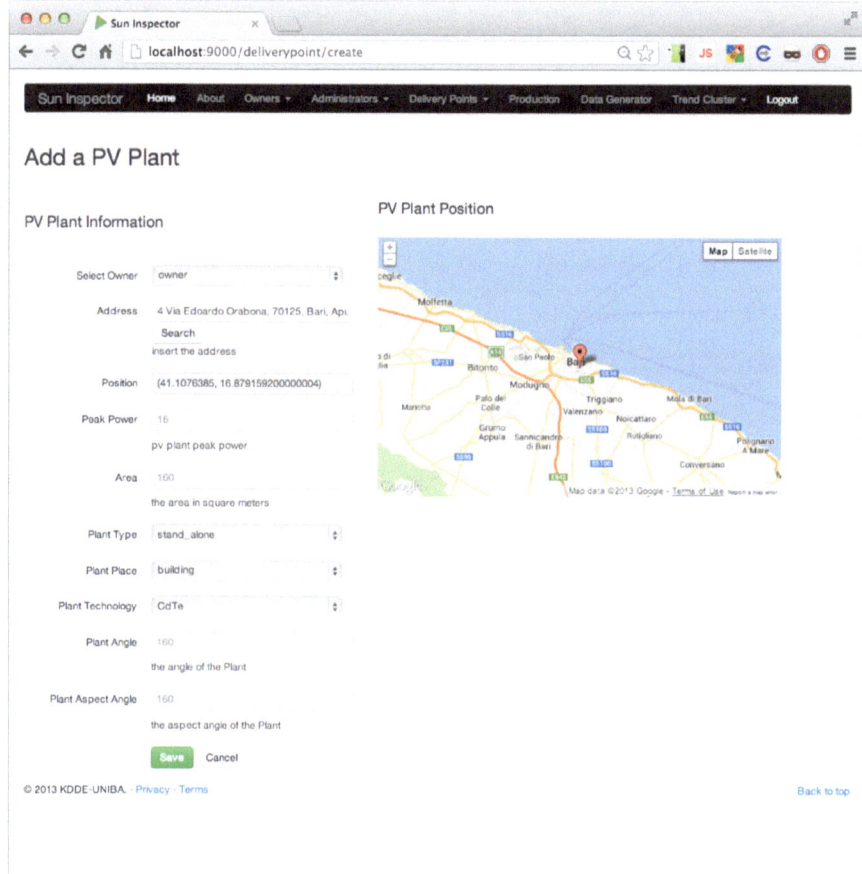

**Fig. 5.4** Sun Inspector web interface: the web page to register a new PV plant

### 5.1.1.3 Trend Cluster-Based Summarization

This component wraps the system SUMATRA, described in Sect. 2.3. Contrary to the original system, where the data windows are consumed from a buffer, the Trend Cluster component loads the windows to be processed from the database. This is done, in order to allow PV customers to check the latest raw productions of their PV plants. The summarization process is implemented as a time-scheduled service. Given the neighborhood distance and the domain similarity threshold parameters, SUMATRA discovers trend clusters by the three-stepped process that:

1. loads the data window from the database;
2. computes the trend clusters of the data window;
3. stores discovered trend clusters in the database.

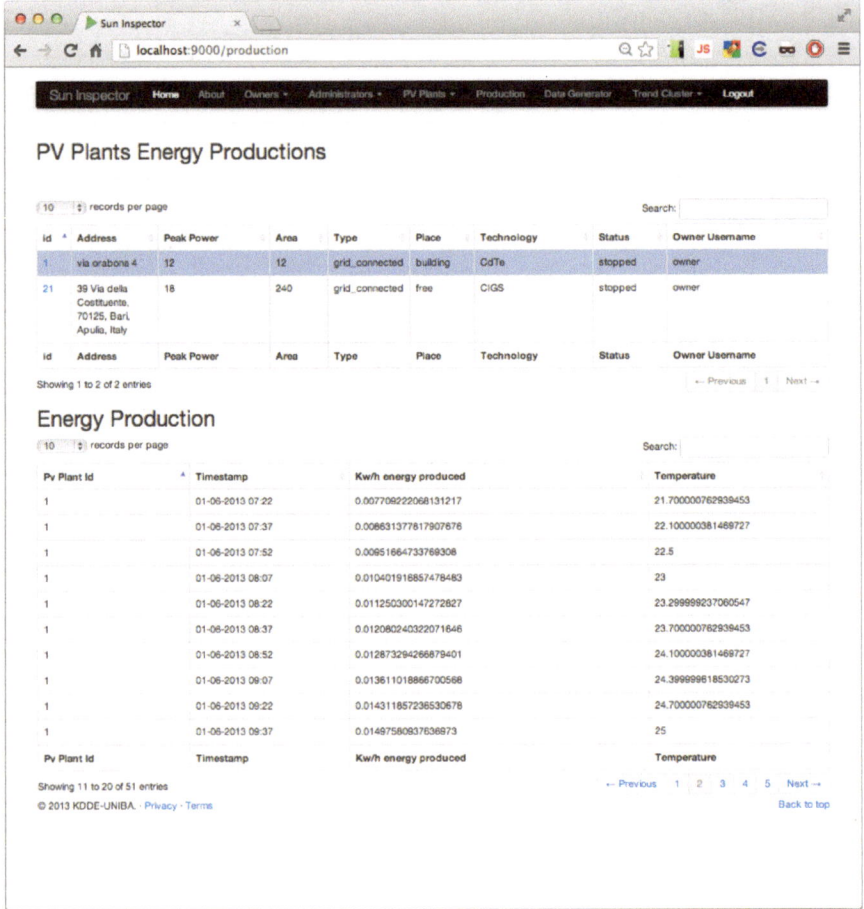

**Fig. 5.5**  Sun Inspector web interface: example of energy production data

The summarization process can be started via Sun Inspector. Figure 5.6 shows the web page to start the trend cluster discovery process. The input parameters include:

1. **Network:** the type of measures to be processed by SUMATRA (e.g., energy production, temperature,…).
2. **Starting Time:** the time to start loading data windows from the database.
3. **Interval snapshot in minutes:** elapsed time (in minutes) between consecutive snapshots, which compose the windows.
4. **Window size:** the number of snapshots in a window.
5. **Minimum threshold:** the domain similarity threshold used to consider PV plant productions as similar.
6. **Max distance:** the neighborhood distance between PV plants to be considered neighbors.

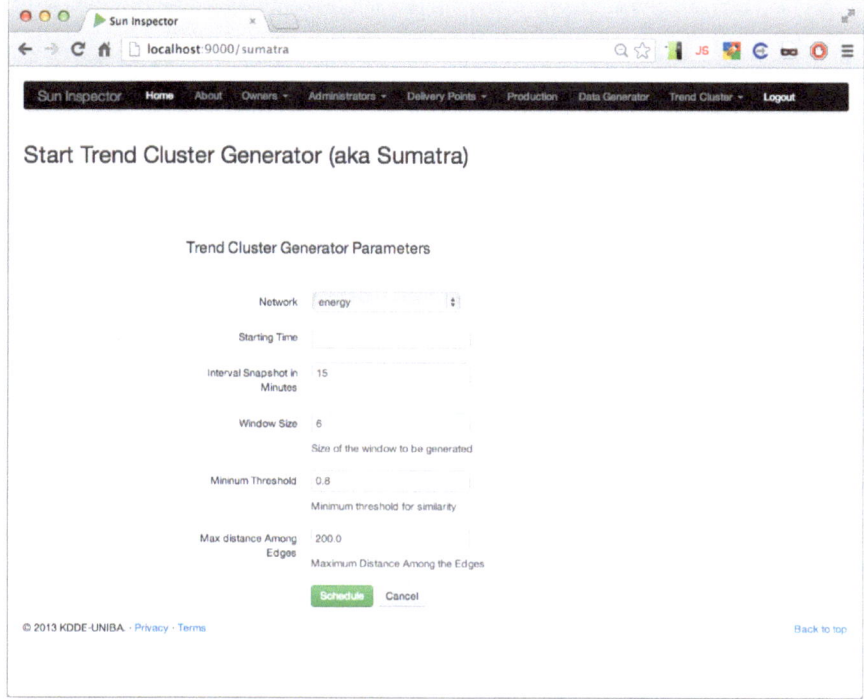

**Fig. 5.6**   Sun Inspector web interface: the web page to start SUMATRA

By pressing the button "schedule", SUMATRA starts to discover trend clusters window-by-window. The computed trend clusters are indexed and made available to end users though the graphical web interface. Users can visualize trend clusters and check trend productions (See Fig. 5.7).

### 5.1.1.4  Data Generator

The Data Generator component is a web service, which allows users to simulate the energy production of a PV plant. It is implemented by wrapping an extension of the web application *Photovoltaic Geographical Information System-Interactive Maps (PVGIS-IM)* implemented by the European Commission. PVGIS-IM[2] is a radiation database, which can be used to estimate the solar electricity produced by a PV plant over the year as well as the monthly/daily solar radiation energy, which hits one square meter in a horizontal plane in one day. It can be queried by filling in a form with several parameters related to the geographic position, the inclination and the orientation of the PV plant. The Data Generator component wraps

---

[2] http://re.jrc.ec.europa.eu/pvgis/apps4/pvest.php

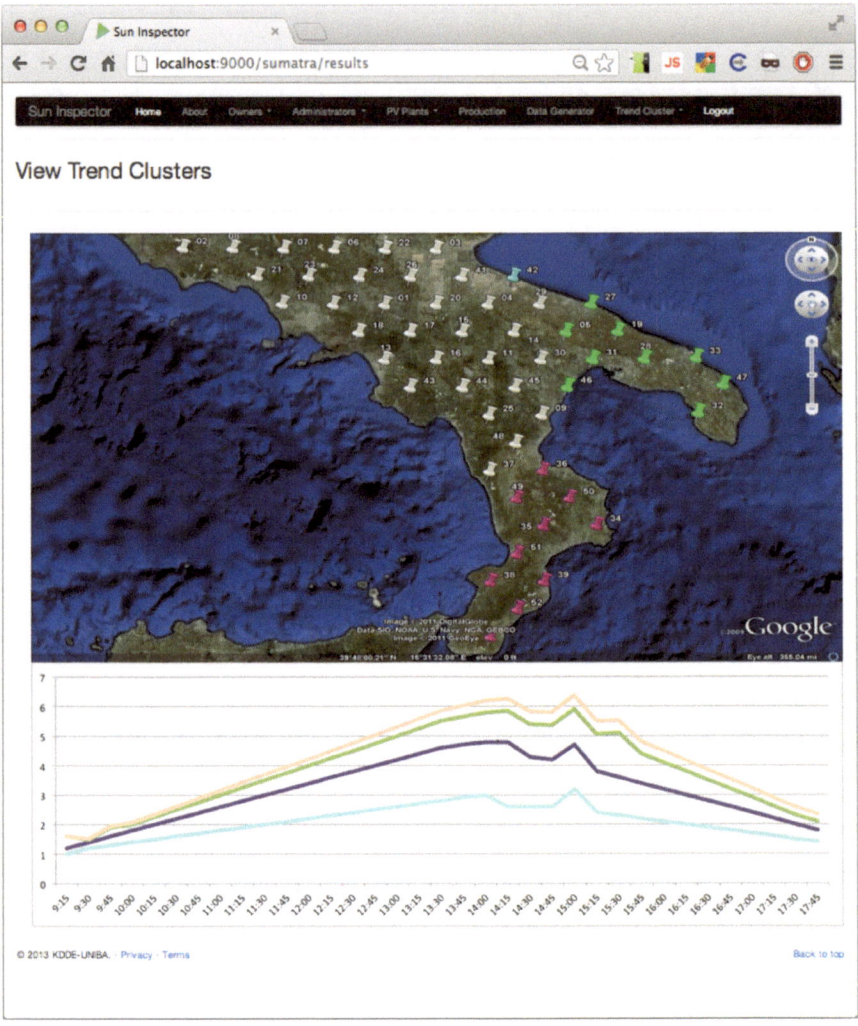

**Fig. 5.7** Sun Inspector web interface: the web page to view trend clusters discovered by SUMATRA

the PVGIS-IM by offering a Rest Web Service interface, which can be queried by automatic services. Moreover, it offers a new service that, given the characteristics of a PV plant as input, simulates its day-by-day energy productions. Figure 5.8 shows the Sun Inspector web page to use the Data Generator. To simulate the PV plant energy productions, the Data Generator combines the solar irradiation queried from PVGIS-IM with the characteristics of the PV plant. The source code of the data generator is available at the following url http://bitbucket.org/kddeuniba/datagenerator.

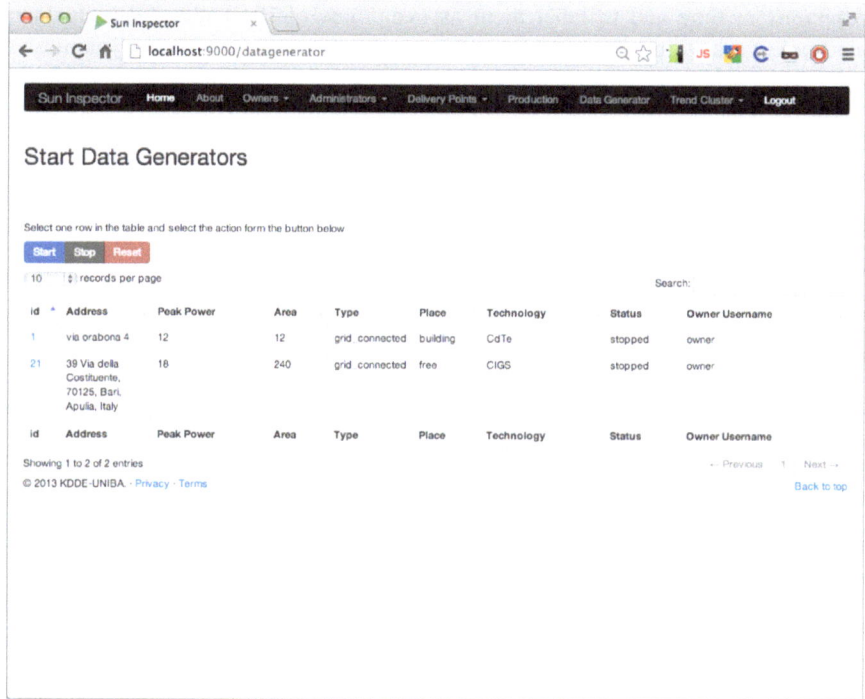

**Fig. 5.8** Sun Inspector web interface: the web page to use the data generator component

## 5.2 Fault Diagnosis in PV Plants: A Data Mining Solution

We describe a fault diagnosis service [2], which makes a network of PV plants smart, by automatically alerting the presence of faulty plants and promptly arranging repair activities.

The scenario that we consider is a network of PV plants, which periodically transmit measurements of the plant energy production to a central server. By considering that the production of electrical energy depends on how much light strikes the station, we have designed a smart monitoring service, which takes into account that the light amount may change with both space (i.e., latitude and longitude of a plant) and time (i.e., the season of the year).

This idea moves away from the plethora of monitoring systems [3–6] already developed by the PV community. In any case, existing systems neither cope with the spatial arrangement of PV plants nor process the produced stream of data along the temporal dimension. On the contrary, we have decided to capitalize on the knowledge which can be extracted by considering the spatiotemporal distribution of the energy production measure. In particular, we have designed a smart fault diagnosis service, which permits the identification of the plant productions which are continuously *suspicious* in time and to label them as symptoms of PV faults. Once again in this book we have used trend clusters to model the spatiotemporal dynamics of data.

The presented fault diagnosis service [7] is decomposed into two sub-services, that is, (1) learning a yearlong model, which describes the expected energy production within the boundary of a fixed region along the time of 1 year; and (2) using this model to determine, in real time, the fault risk of a plant installed anywhere inside the boundary of the region under examination.

### 5.2.1 Model Learning

The energy production model is learned by processing a training set, which collects the periodic measurements of energy production, which are transmitted over the time of 1 year by a training set of PV plants installed in the region under observation. The trend clusters, which are discovered with the sliding window model (Sect. 4.2), define the energy production model of the region.

The learning problem is formally defined as follows.

Given:

1. A network $K$ of training PV plants distributed in the region of analysis.
2. A training yearlong time horizon $T$, which is discretized in $n$ $p$-spaced time points.
3. A series of training data snapshots, which collect the energy productions measured from $K$ at the discrete time points of $T$.

The goal is to learn the *yearlong energy production model* $\mathcal{E}(K, T)$ as a series of $n$ timestamped models of the energy production, one for each time point in $T$,

$$\mathcal{E}(K, T) = \mathcal{E}(K, t_1), \mathcal{E}(K, t_2), \ldots, \mathcal{E}(K, t_n). \tag{5.1}$$

Each model $\mathcal{E}(K, t_1)$ (with $i = 1, 2, \ldots, n$) synthesizes the expected energy production of $K$ at the specific time point $t_i \in T$.

In this study, we have decided to maintain an insight into the historical behavior of each PV plant and take advantage of this insight in the fault risk evaluation. Therefore, the trend clusters, discovered in the training set with a sliding window model (Sect. 4.2), are used to represent the energy production model. This means that, for each time point $t_i$, $\mathcal{E}_i(K)$ is the set of trend clusters of the training set, which are labeled with the time horizon $t_{i-w+1} \rightarrow t_i$. To be able to compute this model for every time point of $T$, we consider $T$ as a circular list, so that $t_k$ is treated as the predecessor of $t_1$ (and vice-versa, $t_1$ is treated as the successor of $t_k$). The size $w$ of the sliding window model represents the size of the memory of the model.

## 5.2.2 Fault Detection

The yearlong energy production model $\mathscr{E}(K, T)$ is used to monitor the efficiency of every plant, which is installed in the region surrounding the training network $K$. At each time point, the set of trend clusters, associated to the corresponding sliding window, is selected from the energy production model. Then the areal unit (spatial cluster), which contains the monitored plant, is identified and the trend polyline time series associated to this cluster is compared with the time series of the real variation of energy productions observed for the plant over the recent window. The dissimilarity between these two time series is computed to estimate the degree of fault risk.

The fault risk detection task is formally formulated as follows.

Given:

1. A yearlong energy production model $\mathscr{E}(K, T)$.
2. A PV plant $k$ that continuously transmits periodic measures of the energy production at $p$-spaced consecutive time points.
3. A certain time point $t_i$.

The goal is to measure the fault risk degree $f_R(k, t_i)$ of the plant $k$ at the specific time point $t_i$ and raise an alarm when the computed degree goes over a user defined threshold.

The fault risk degree is estimated by computing the dissimilarity between the observed series $kZ$ of energy production measurements, produced by the plant $k$, and the expected measurements $eZ$ of the same plant for the window with time horizon between $t_{i-w+1}$ and $t_i$. Our motivation for evaluating the observed/expected values over a window, rather than at a single time point, is that we intend to detect the plant whose energy production is persistently anomalous along a time horizon. In this way, we can filter out noise, which may affect data, and reduce false alarms.

To illustrate how $f_R(\cdot, \cdot)$ is computed, we first specify how the observed series and expected series are obtained and then we explain how dissimilarity between these data is computed and used to estimate the fault risk degree.

Observed data

The observed data for the plant $k$ at the time $t_i$ are the series of the most recent income $w$ measures of energy production produced from $k$. Formally, let $Z$ be the energy production variable, so we have that:

$$kZ(k, t_i) = z(k, t_{i-w+1}), z(k, t_{i-w+2}), \ldots, z(k, t_{i-1}), z(k, t_i). \tag{5.2}$$

For each monitored plant $k$, when a new data snapshot is produced in the monitored network, the oldest energy production measure is discarded from $kZ(k, t_i)$ (sliding data), while the new measure is added to $kZ(k, t_i)$.

Expected Data

Let $\hat{t}_i$ be the time point of $T$ which is closest to $t_i$ (regardless of the year). Then $\mathcal{E}(K, \hat{t}_i)$ is the expected model of the energy production of $k$ at the time $t_i$. This model is recovered from $\mathcal{E}(K, \hat{t}_i)$ by identifying the cluster $\mathcal{C}$, which hosts the majority of training neighbors of $k$ and returning the $w$-sized trend polyline time series $\mathcal{L}$, which is associated to $\mathcal{C}$. Let $(\hat{t}_{i-w+1} \rightarrow, \hat{t}_i, \mathcal{C}, \mathcal{L})$ be the selected trend cluster, so we have that:

$$eZ(k, t_i) = \mathcal{L}[\hat{t}_{i-w+1}], \mathcal{L}[\hat{t}_{i-w+2}], \dots \mathcal{L}[\hat{t}_{i-1}], \mathcal{L}[\hat{t}_i]. \tag{5.3}$$

Fault Risk Degree Computation

The fault risk degree $f_d(\cdot, \cdot)$ is computed as follows:

$$f_d(k, t_i) = d(kZ(k, t_i), eZ(k, t_i)) = \tag{5.4}$$

$$= \frac{\displaystyle\sum_{j=i-w+1}^{i} diss(z(k, t_j) - \mathcal{L}[\hat{t}_j])}{w}; \tag{5.5}$$

where $diss(\cdot, \cdot)$ is computed as follows:

$$diss(v1, v2) = \begin{cases} 1 & \text{iff } \|v1 - v2\| \geq \delta \\ 0 & \text{otherwise} \end{cases}, \tag{5.6}$$

and $\delta$ is the trend similarity threshold according to which trend clusters are computed.

Here $f_d(k, t_i)$ can range between zero (i.e., the observed value is persistently similar to the expected one over the time horizon of the entire window) and one (i.e., the observed value is dissimilar from the expected one in one or more time points of the window). The higher $f_d(k, t_i)$, the higher the fault risk.

### 5.2.3 A case Study

We present an application, where we monitor PV plants distributed in the South of Italy, which weekly (p=1 week) produce measurements of total energy productions (in kw/h). A description of these data is reported in Sect. 2.5.5.

We consider 52 training PV plants in the South of Italy, distributed as shown in Fig. 5.9a. Each training plant is 0.5 degrees in latitude and 0.5 degrees in longitude apart the others (see the white pushpins in Fig. 5.9a. A yearlong production model is learned with sliding window size $w = 8$ and domain similarity threshold

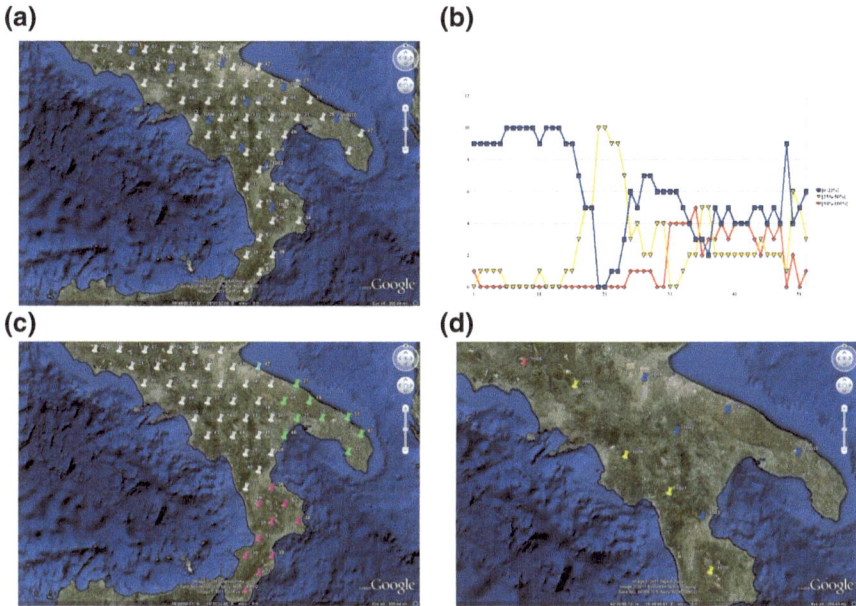

**Fig. 5.9** **a** Training PV plants (*white pushpins*) and testing PV plants (*blue pushpins*). **b** Number of plants weekly classified into a low risk zone (*blue*), medium risk zone (*yellow*), and high risk zone (*red*). The trend cluster partition of the South of Italy territory **c** and the fault-based coloring **d** of the testing PV plants as it appeared at the 26th week of the testing monitoring activity

$\delta = 1.5\,kWh$. This model is learned off-line from yearlong training data. The model is then used to monitor on-line 10 testing PV plants, which are installed randomly in the South of Italy (see the blue pushpins in Fig. 5.9a. The energy production measures of the testing PV plants are generated with PVGIS (http://re.jrc.ec.europa.eu/pvgis/), but testing data are perturbed with randomly added noise.

The fault risk degree, computed week-by-week, is visualized on the map. Plants are colored on the basis of the fault risk degree, so that the plant visualization is updated accordingly. For this study, we have assigned a color to three zones of risk, that is, *low fault risk zone* (blue), where the risk degree is less than 0.25, *medium fault risk zone* (yellow), where the risk degree is between 0.25 and 0.5 and *high fault risk zone* (red), where the risk degree is greater than 0.50.

The number of testing plants predicted in each risk zone is plotted in Fig. 5.9b. An example insight into the fault risk computed in the 26th week of the monitored year is reported in Figs. 5.9c, d. In particular, Fig. 5.9c shows the South Italy partitioning on the basis of trend clusters, while Fig. 5.9d plots the fault risk computed for each testing plant. Plants are colored on the basis of the computed fault risk and alarms are raised in correspondence to the high risk faults. Alarms are always raised in correspondence to perturbed measurements, which exhibit the typical characteristics of a fault scenario.

## 5.3 Summary

In this chapter, we have illustrated two applications of sensor data analysis in the specific context of smart networks of PhotoVoltaic plants. The former is a business intelligence solution to monitor the efficiency of PV plants. The latter is a fault diagnosis service which resorts to trend cluster discovery to monitor the energy production of a network of PV plants and raise an alarm in the presence of faults.

## References

1. C. Pautasso, O. Zimmermann, F. Leymann, in *Restful web services versus big web services: making the right architectural decision*. Proceedings of the 17th International Conference on World Wide Web, WWW '08, (ACM, New York, 2008), pp. 805–814
2. S. Ding, *Model-based Fault Diagnosis Techniques* (Springer, New York, 2008)
3. M. Zahran, Y. Atia, A. Alhosseen, I. El-Sayed, Wired and wireless remote control of PV system. WSEAS Trans. Syst. Control **5**, 656–666 (2010)
4. M. Zahran, Y. Atia, A. Al-Hussain, I. El-Sayed, *Labview based monitoring system applied for pv power station*. In International Conference on Automatic Control, Modeling and Simulation, ACMOS 2010, (WSEAS, Stevens Point, Wisconsin, USA, 2010). pp. 65–70
5. V. D. Ulieru, C. Cepisca, T. D. Ivanovici, A. Pohoata, A. Husu, L. Pascale, Measurement and analysis in PV systems. In International Conference on Circuits, ICC 2010, (WSEAS, Stevens Point, Wisconsin, USA, 2010) pp. 137–142
6. T. Sugiura, T. Yamada, H. Nakamura, M. Umeya, K. Sakuta, K. Kurokawa, Measurements, analyses and evaluation of residential PV systems by Japanese monitoring program. Sol. Energy Mater. Sol. Cells **75**(3–4), 767–779 (2003)
7. A. Ciampi, A. Appice, D. Malerba, A. Muolo, in *An Intelligent System for Real Time Fault Detection in PV Plants*. Smart Innovation, Systems and Technologies, vol 12 (2012), pp. 235–244, doi:10.1007/978-3-642-27509-8_19

# Glossary

$b$ Influence boundary of the interpolation sphere
$\mathscr{C}$ Cluster
**DFT** Discrete Fourier Transform.
**DHW** Discrete Haar Wavelet
$\mathscr{H}(space(T))$ Space hierarchy
$\mathscr{H}(T)$ Time hierarchy
**IDW** Inverse Distance Weighting
$K$ Geosensor network
**lower case u, v** Sensor sources
**PVP** PhotoVoltaic Plant
$\mathscr{Q}$ Geodata cube
**rmse** Root mean square error
**se** Stability error
$size^{\%}$ Compression size
$T$ Time line
$w$ Window size
$z_t()$ Field function
$z(T, K)$ Geodata stream
$Z$ Geo-physical field (variable)
$\mathscr{Z}$ Trend polyline
$|\cdot|$ Cardinality of a set
$\delta$ Domain similarity threshold
$\varepsilon$ Compression error threshold
$\gamma(h)$ Sample variogram
$\eta$ Variogram nugget
$\iota(t)$ Forecasting model intercept

A. Appice et al., *Data Mining Techniques in Sensor Networks*,
SpringerBriefs in Computer Science, DOI: 10.1007/978-1-4471-5454-9,
© The Author(s) 2014

$\nu(t)$ Forecasting model seasonality
$\rho$ Variogram range
$\sigma$ Compression degree threshold
$\varsigma^2$ Variogram sill
$\tau(t)$ Forecasting model intercept

# Index

A. Appice et al., *Data Mining Techniques in Sensor Networks*,
SpringerBriefs in Computer Science, DOI: 10.1007/978-1-4471-5454-9,
© The Author(s) 2014